LOGIC MADE EASY

R.H. WARRING

TAB BOOKS Inc.
Blue Ridge Summit, PA

Published in the U.S.A. in 1985 by TAB BOOKS Inc.
Printed in the United States of America
First published 1984 by Lutterworth Press.
Reprinted by permission of Lutterworth Press.

FIRST EDITION

THIRD PRINTING

Library of Congress Cataloging in Publication Data

Warring, R.H. (Ronald Horace), 1920-
   Logic made easy.

   Includes index.
   1. Logic, Symbolic and mathematical.   I. Title.
QA9.W375   1985   511.3   85-12578
ISBN 0-8306-1853-8 (pbk.)

TAB BOOKS Inc. offers software for sale. For information and a catalog, please contact TAB Software Department, Blue Ridge Summit, PA 17294-0850.

Questions regarding the content of this book
should be addressed to:

Reader Inquiry Branch
TAB BOOKS Inc.
Blue Ridge Summit, PA 17294-0214

Contents

Introduction

'Contrariwise', continued Tweedledee, 'If it was so, it might be; and if it were so, it would be; but as it isn't, it ain't. That's logic.'

from Lewis Carroll's *Through the Looking Glass*

That, as a concise explanation of logic, is quite clever. But Lewis Carroll was not only an outstanding writer; he was also a mathematician and an active practitioner of logic, or logician. The dictionary defines logic much more broadly as the science of reasoning, proof, thinking or inference – which is not too logical when you come to look into it. It merely hints at the various possible forms of logic, which really start with commonsense. That is the sort of logic most people can understand, but it is not *true* logic. True logic follows rules which must not be broken to justify *logical* conclusions or answers.

The first thing to observe when looking at the subject of logic in depth is that there are many different *types* of logic – so there is hope for everyone to be able to find a particular type of logic which he or she can readily understand and apply to reach logical conclusions or answers. This type of logic will then become a useful, everyday tool – much more effective than mere commonsense.

This book has been planned with that in mind. It has been divided into chapters which deal with different types of logic. Some will be easy to understand; others possibly quite obscure at first. In the latter case, skip those for a start. Read only those which immediately 'make sense'. You can then start to apply that type of logic right away. Go back to the others later on, when you will discover that there is an inter-relationship between all the different types of true logic. Logic is not just a subject for philosophers or mathematicians – it is something everyone can use and benefit from.

Logic today, in fact, is a 'popular' subject. There are various journals and publications, for instance, devoted solely to presenting puzzles and problems for solution by 'logic' as a stimulating mental exercise or even hobby-interest – just as crossword puzzles attract their addicts. The same sort of 'logic' problems have also been set for years in aptitude tests. Largely, however, this is a back-to-back form of logic. The originator of the problem starts with the answer and derives the facts or clues to fit it, which are then presented to the reader to solve and produce the *same* answer.

Logic, too, has become a powerful modern design tool, particularly for the complex circuits used in microprocessors and computers; the more complicated 'engineering' control circuits, etc; and, of course, computer-aided design. All the modern forms of practical (mathematical) logic deal in binary relationships, a tailor made language for computers and the ever-expanding field of *design by logic* rather than by combersome equations and tedious mathematical solutions. At the same time binary language, as expressed in Truth Tables and Boolean algebra, is an extremely simple, useful tool for application to problems in 'ordinary' logic, where it can often be a much more compact (and exact) alternative to deductive logic. The basic mathematical principles involved date back for more than a century. Before they had a definite practical application they remained of interest only to the 'pure' logician concerned with academic studies. Bertrand Russell – a leading philosopher/mathematician – even gave up his studies of mathematical logic because the solutions obtained were 'too exact'.

On the other hand, *formal* logic has been the delight of philosophers for two thousand years, with philosophers greatly outnumbering mathematicians throughout. This has produced the logicians whose proper business, as defined by a leading (American) authority is 'the investigation and formulation of general principles concerning what follows from what; and whether particular examples of his own reasoning are valid or not is irrelevant ... Correct reasoning, however praiseworthy, does not itself contribute to logic.' The first sentence of this virtually repeats Tweedledee's definition of logic. The remainder contains two statements which may seem surprising. 'Correct' logic does not have to be valid (i.e. the reasoning behind it does not necessarily have to be true!). Also if the reasoning *is* correct, it is not necessarily logic! You will find these two apparent contradictions of the meaning of logic explained in the chapter on Deductive Logic.

Historically, the origins of formal logic date back to the Greek philosopher Aristotle (384–322 BC), who developed the basic theory of syllogism, which has been the core of deductive logic ever since as first developed by the Peripatetic school. Alternative approaches to formal logic were subsequently developed by other groups of philosophers, of which probably Stoicism (Stoic logic) is the most outstanding as the source of sentential calculus. Except for the dedicated student of philosophy, then there is little else of note until the middle of the sixteenth century when the first of the 'mathematical logicians' began to appear, culminating in the development of Boolean algebra almost simultaneously by George Boole (1815-64) and Augustus De Morgan (1806-71). Mathematical logic, in its modern form, is finally accredited to Gottlob Frege (1848-1925) with his consistent and complete development of the sentential calculus; subsequently further developed by Bertrand Russel and Alfred North Whitehead in *Principia Mathematica*.

All this is concerned, basically, with the logic used by philosophers (and of necessity has omitted many other important names in the development of philosophical logic). Simplifying it to a degree, Aristotle remains the inventor and inspiration behind deductive logic, as it is still used today. Mathematical logic based around sentential calculus, was 'finalized', as it were, only at the beginning of the present century and because of its complexity demanded a new artificial language to support it. It thus remains largely the tool of philosophers.

Indeed, as far as philosophers are concerned, the algebraic approach to logic using binary relationships remained as a weaker tool, although as noted previously it has now become an important design tool. For that reason, if you go to your local library to find books on logic, virtually all will be catalogued and displayed under Philosophy. Treatment of practical modern logic will only be found in parts of books on the Electrical and Mathematical shelves.

It is to be hoped, therefore, that this present book will fill a real need for introductory coverage of *all* types of logic in a single volume; and above all make these types of logic easy to understand and apply.

CHAPTER 1

Some Types of Logic

Commonsense Logic

Commonsense logic is deriving conclusions from personal experience and/or knowledge. A conclusion that something makes sense, so it is right; or something does not make sense, so it is wrong. Commonsense does not necessarily produce correct answers, however; and is not necessarily 'logical' at all – especially when compared with deductive or mathematical logic, for example; or even established facts.

Commonsense 'logic', for example, would maintain that brick, stone, metals are all hard and solid substances. Science establishes that the atomic structure of any solid substance is almost entirely empty space. Commonsense finds it difficult, or impossible, to accept such a fact. It is not understandable, so it is not real. Even less 'realistic' is time-dilation in space travel; or the quantum theory which holds that *everything* can be reduced to and analysed in terms of wave forms.

Yet commonsense is the logic most individuals use for solving ordinary, everyday problems. Applied to pure logic problems it can even provide instant answers which are *known* to be correct, ignoring any rules of logic and avoiding any necessity of positive *proof* that the conclusion is correct.

A person familiar with philosophical logic and deduction would immediately identify the following as a syllogism with an invalid conclusion:

> All dogs are animals
> All cats are animals
> Therefore all cats are dogs.

He could further go on to *prove* that this is an invalid syllogism by drawing a Venn diagram, or by showing that it breaks one or more of the *rules* of deductive logic. Equally, he would probably be prepared to argue this proof at some length.

Commonsense logic has not heard of syllogisms or valid or invalid arguments. It simply affirms, without argument, that cats are *not* dogs.

To do this, however, it has to *know* there is a difference between cats and dogs. Without any knowledge at all of the French language, for example, commonsense logic could also maintain that all cats are chats is also wrong (because the spelling is wrong).

This, incidentally, leads to an interesting digression.

All cats are animals
All chats are animals
Therefore: (i)   all chats are cats
            (ii)   all cats are chats.

True or false? (i) is true and (ii) is false. But this is a question of *knowledge*, not logic. (In the case of (ii) all cats are not male (French) cats.)

Here is a classic problem in commonsense logic. A bear walked one mile due South, then turned to the left and walked one mile due East. Then it turned to the left again and walked one mile due North and arrived back at its starting point. . . . What was the colour of the bear?

Now apparently the bear walked three legs of a square, like this ⌐ . But since it ended up where it started from its actual path

must have been a triangle △. The only two places in the world where this could happen is if the starting point is either the North Pole or the South Pole. The South Pole is ruled out since it is impossible to travel South from it. So the bear was at the North Pole, i.e. it was a polar bear. So the bear was *white*.

Note again that this problem is solved by *knowledge*, not logic as such – although it needs a logical type of mind to apply that knowledge to a particular problem. So commonsense logic can perhaps best be described as *logical reasoning* based on the available facts, and drawing on additional knowledge or experience to arrive at an answer. It could almost – but not quite – be called inductive logic, which is a recognized category. It is certainly *not* true deductive logic, which does not permit argument *outside* the facts available.

In this particular example, too, logical reasoning can also be *proved* by geographical fact – except for one thing. There is the remote possibility that a *brown* bear could have been taken to the North Pole by aircraft, say, as an environmental experiment.

Sherlock Holmes's Logic

Almost everyone must be familiar with the infallible reasoning of Sherlock Holmes – his ability to put together a complete picture from the most meagre of clues whilst the amiable Watson was simply confused by the situation. Deductive logic at its best – or is it?

In fact it is not, although it is based on the principle of deductive logic. It is fiction, written 'backwards' from the answer. Clever, imaginative writing where the answer (conclusion) is first established, the clues (premises) then extracted from the answer and then whittled down to the absolute minimum to be ultimately acceptable for justifying the answer. The possibility of alternative answers is not considered for it does not fit in with the story, or the character of Sherlock Holmes.

Let's face a simple situation which Sherlock Holmes' type logic would answer immediately with ease.

One Spring day you come across two round stones and a carrot lying close together on a grass verge. What do you deduce from that?

The Sherlock Holmes type of mind would immediately answer: 'A boy built a snowman there in the second week of February.' (Sherlock Holmes himself would probably have gone on to describe the boy in more detail; where he went on holiday last (by the type of stones); and even the snowman itself).

Now possibly this answer is right. Why a snowman? Because the stones were used for eyes, and the carrot for a nose. Why a boy? Because a girl would have taken more trouble and moulded the nose in snow. Why the second week in February? Beacuse that was when there was the last fall of snow that laid heavily.

None of these conclusions is supported by fact, so it is difficult to justify them on any *logical* basis. Indeed there are many other possible answers, for example:

(i) The two stones just happened to be there, anyway. The carrot fell out of someone's shopping bag at that particular spot.

(ii) Digging up his vegetable plot, a gardener turned up two stones and an old carrot. He threw them all over the hedge, where they landed on the grass verge close together. Neither very interesting, but both equally as plausible as the 'snowman' theory, or even more so.

(iii) One night a burglar approached the house the other side of the grass verge carrying two stones he could use to break a window and force an entry. He thought the house was empty, but it was not; so he turned, ran and jumped over the hedge on to the grass verge, dropping the stones where he landed. The woman who was in the house and saw him was in the kitchen. She picked up the nearest object, which happened to be a carrot and threw it at the burglar. The carrot landed in the same spot as the stones.

That answer has 'written a story' around two simple facts. See how many other different stories, or answers, you can think of.

Ordinary Language

Arguments in ordinary language are often difficult to *prove*, even though they may lead to correct conclusions. This is the basic weakness of ordinary language for dealing with problems in logic. Where proof of the validity or otherwise of the conclusion is necessary, the available information may first need changing into what is known as a standard-form categorical syllogism before it can be analysed fully under the rules of *deductive* logic. The term 'standard-form categorical syllogism' is extremely off-putting when first met, but basically means reducing the component propositions into three separate terms. The validity (or otherwise) of the argument is then simple to establish (*see* chapter on Deductive Logic).

As an example, consider the following forms:

No wealthy persons are vagrants
All barristers are rich people
Therefore, no high court advocates are tramps

These contain six terms, but in fact this is because of synonymous descriptions. Thus 'wealthy people' and 'rich people' are the same; so are 'barristers' and 'high court advocates'; and so are 'vagrants and tramps'. Thus, still using ordinary language, here is the same thing in *standard-form* containing just *three* terms.

No rich people are vagrants
All barristers are rich people
Therefore, no barristers are vagrants.

That is now a standard-form categorical syllogism, which is easy to prove as valid argument.

At the same time this demonstrates a basic rule applicable to all types of logic. The information *necessary* to be able to solve the problem, or come to a conclusion, can at first appear obscure because it is wrapped up in a lot of ordinary language. The 'facts' of the question have to be extracted and put down in their simplest form, eliminating any ambiguity of duplication (synonymous description).

Reductio ad absurdum

Reductio ad absurdum is a particular type of logic favoured by philosophers, as well as being applicable to other types of logic. It means, quite simply, showing that an assumption is false by deriving a further conclusion from it which is absurd (i.e. cannot be true). In fact, it can be argued that *reductio ad absurdum* itself is absurd as applied in deductive logic since it argues at length about things which are *not* true. Debating the question rather than seeking the answer. In mathematical logic, however, *reductio ad absurdum* can be quite precise. It can be used to *prove* that something is impossible, i.e. that a particular answer is not a correct solution or method of solution; or that there is no answer to that problem.

Suppose the problem is to complete a series of numbers using each of the ten digits 0,1,2,3,4,5,6,7,8 and 9 once and once only so that the sum of such numbers composed is 100. Can it be done? The *reductio ad absurdum* approach will show that it cannot.

Using the digits as they stand, added together

$$0+1+2+3+4+5+6+7+8+9=45$$

This is well short of the total required, so some of the numbers will have to be double figures instead of single figures to make up the additional amount in *tens*. Call this sum T for tens. The sum of the remaining figures is then $45-T$, so we get the following equation expressing the requirement

$$10 \times T + (45 - T) = 100$$

Solving this gives $T = 55/9 = 6.1111\ldots$.

This is obviously absurd. You cannot have 6.111... digits denoting tens. T *must* be a whole number. So, by *reductio ad absurdum*, the problem is *not* solvable.

By all means apply *reductio ad absurdum* to deductive logic as well - but be prepared for a lengthy debate if discussing it with someone else. In philosophical logic there can be different *opinions* as to what is absurd or not!

Indirect Proof

Indirect proof is closely related to *reductio ad absurdum*, but it works in the opposite sense. It 'proves' a deduction or assertion by establishing that the *opposite* assumption is false. A handy trick for politicians, this, to prove their policies right by expounding at length on how the Opposition is wrong.

Even more so than *reductio ad absurdum*, indirect proof is open to objections when applied to philosophical logic, but again in mathematical logic it can be quite positive.

Here is a simple example. If we add all the numbers possible together (i.e. $1+2+3+4+5+$ etc., etc.) we conclude logically that there is no end to such a series and so the sum of all possible numbers is an infinitely large quantity – normally called infinity, or designated ∞. Thus $1+2+3+4+5+$ etc., etc. $=\infty$

Suppose, now, we adopt the *opposite* assumption that there *is* an end to such a series. If that is so, a further 1 can be added to it to give:

$$(1+2+3+4+5+ \text{ etc., etc.}) + 1 = \infty + 1$$

But there cannot be a quantity $\infty + 1$ because ∞ (infinity) is already an infinitely large number (there cannot be a larger number). So $\infty + 1$ is impossible. Thus this second (opposite) assumption is false. Hence the original assumption must be true.

'Proving' that $2 = 1$

There are quite a number of ways of 'proving' mathematically that $2 = 1$. Indirect proof immediately asserts that any such 'proof' is false. Sometimes this can be shown directly, i.e. *proved* false, without having to look too far. The trick of introducing $\infty + 1$ in an equation is one common method of 'proving' $2 = 1$. In some cases, though, it can be difficult to see *where* the 'proof' is wrong.

Take the following series of numbers added together, for example, calling the sum X:

$$X = 1 - \tfrac{1}{2} + \tfrac{1}{3} - \tfrac{1}{4} + \tfrac{1}{5} - \tfrac{1}{6} + \text{ and so on indefinitely.}$$

Now multiply by 2 to give a second series

$$2X = 2 \times 1 - 2 \times \tfrac{1}{2} + 2 \times \tfrac{1}{3} - 2 \times \tfrac{1}{4} + 2 \times \tfrac{1}{5} - 2 \times \tfrac{1}{6} + \dots\dots\dots\dots$$

which gives

$$2X = 2 - 1 + \tfrac{2}{3} - \tfrac{1}{2} + \tfrac{2}{5} - \tfrac{1}{3} + \dots\dots\dots\dots\dots\dots\dots\dots\dots$$

which can be grouped like this

$$2X = (2-1) + (\tfrac{2}{3}-\tfrac{1}{3}) - \tfrac{1}{2} + (\tfrac{2}{5}-\tfrac{1}{5}) - \ldots\ldots\ldots\ldots\ldots\ldots\ldots\ldots\ldots$$
$$= 1 - \tfrac{1}{2} + \tfrac{1}{3} - \tfrac{1}{4} + \tfrac{1}{5} - \tfrac{1}{6}\ldots. \text{ and so on } \ldots\ldots\ldots\ldots\ldots\ldots\ldots$$

which is back to the original series, thus

$$2X = X$$
or $2 = 1$

Proving that this mathematical reasoning is wrong can be quite a problem!

No – the fallacy of this proof is not in grouping alternate pairs of fractions and simplifying each group to a single fraction. Since the series extends to an infinite number of fractions, every fraction we 'borrow' from farther on in the series does *not* reduce the number of fractions in the series – nor does it extend infinity beyond infinity.

Lateral Thinking

Lateral thinking is another form of mental discipline which has received considerable recent exposure via Dr de Bono's books – and particularly his TV series on the subject. Loosely, it is the art of thinking problems through in more than one direction at a time. Another form of logical argument, in fact, but only marginally acceptable as a form of logic.

Probably the basic principle involved is best illustrated by example, where something is being *designed*. The designer develops the original idea through his own knowledge and experience into something final, which is then built or produced. To complete it, it utilizes 'standard' parts bought in from another company, variation in performance or quality of which may affect the performance of the final product. Unless the designer has indulged in some degree of 'lateral thinking' to take into account such possible effects (which were outside his immediate design problem), the design could prove unsatisfactory.

Here is a very simple example of this. The call is for a bracket to be designed to hold two parts at a V-angle when screwed or bolted up to the bracket. Accordingly the design office drew up a V-shaped bracket, and somebody in the workshop makes it. Then they find that they cannot possibly get long enough screws or bolts into the holes near the bottom of the V-bracket. So a little 'lateral thinking' at the design stage would have taken into account that not only was the shape of the bracket important, but the fact that it had to be *fitted* with bolts. Too bad if they had produced hundreds of brackets before someone found that they were not usable!

Heuristic Reasoning

Heuristic is the name given to a certain form of logic (or possibly more correctly philosophy) involving a study of the methods and rules of discovery and invention. Put in everyday language it could be described as arriving at a *plausible guess* to the answer of a problem. Heuristic reasoning is not regarded as taut (i.e. exact), but somewhat provisional. It may (that is, the answer) be correct or incorrect. Equally, it may be based on inductive logic, or analogy. In either case it can be a useful mental exercise, but heuristic reasoning is not acceptable as positive or absolute proof.

Analogy

Analogy can be defined as a branch of heuristic reasoning and defined simply as 'a sort of similarity'. From such similarity it is possible to make an educated guess at possible unknowns. However the application of analagous argument is much broader than simple argument. Analogies can be vague, or clear cut. They can be applied to mathematics, philosophy or other forms of argument. In many cases analogous argument and deduction is the only answer to lack of knowledge in tackling a problem. Probably the clearest example here is the case of a mathematical problem where a solution is quite straightforward (and exact) using calculus, but the person's knowledge of mathematics does not extend to calculus. Using analagous reasoning he (or she) could well derive an acceptable answer, if not necessarily exact.

That, in fact, is the weakness of analogy. It is never exact argument and deduction – it is heuristic reasoning.

Logic Languages

One of the basic difficulties in developing an approach to logic is the limitation of ordinary language when it comes to presenting premises and conclusion in a *formalized* manner. (Exactly the same problem arises with computers. You cannot instruct or 'talk' to computers in ordinary language – it has to be a *formalized* language consistent with the input/output capabilities of the computer). So formalizing or adapting the style of natural language has been used by logicians since the earliest days, although the first successful attempts to produce a complete artificial language on this basis is due to the German logician Gottlob Frege and only appeared towards the end of the last century. The idea has since been developed extensively, replacing words with symbols, and developing theorems of logic.

The aim is very simple – to give each expression an exact meaning, free from context (which can often confuse or produce ambiguity of meaning in a natural language); and manipulate such expressions in a logical manner (determined by rules and theorems). In practice this becomes a vast subject on its own, far beyond the scope of this book, and so is only mentioned as such. After all, the title is *Logic Made Easy*, and translating a natural language into an entirely new artificial language is not easy. Just as it is not easy to learn computer language, although with the great demand for such knowledge there are now a considerable number of simplified computer languages which are (relatively) easy to learn. Millions of people now use computers. The greatest number of computers in use, indeed, are the home computer types. The number of people seriously interested in a comprehensive formalized logic language probably barely runs into thousands. Hence this type of language remains largely the prerogative of academics.

For instance, you start by translating a simple premise:

some A are B

into something like:

$(\exists x)(Px \mathbin{\&} Ox)$

and a little later on may find you have produced an expression something like this!

$(x)(y)(P_x \text{ ad } O_y) \mathbin{\&} (Sxa_2 \vee Sya_e x) \rightarrow$
$(\exists z)(\exists w)((Lxz \mathbin{\&} Lyw) \mathbin{\&} (Pz \mathbin{\&} Pw) \mathbin{\&} (Sza_2 w \vee Swa_2 z))$*

Nevertheless, you will find that this book does make some considerable use of artificial language using symbols, but in a very, very much simpler form.

---

* Incidentally, the natural language translation of this is – 'for every pair of prime numbers differing by 2, there is a pair of greater prime numbers differing by 2'. There are much simpler ways of expressing the same thing!

Sentential Calculus

For the same reasons that the academic formalized language of
logic is only briefly mentioned, *sentential calculus* is also glossed over. It
relates to the artificial language and, broadly speaking, is based on a
series of *metatheorems* or proofs of general principles characterizing the
language. Every sentence consistent with a metatheorem is then either
a *tautology* (i.e. completely unambiguous) or a generalization of a
tautology.

To whet your appetite – or put you completely off the subject – here
is the start of metatheorem 110 in sentential calculus:

$$\vdash (-\phi \rightarrow \psi) \rightarrow (-\psi \rightarrow \phi)$$

If it looks interesting you can gain some further encouragement from
the fact that there are only 22 theorems in the 100-series covering the
sequential calculus!

CHAPTER 2

Which Type of Logic to Use?

Everyone is born with the inherent ability to do some things well and also to suffer a sort of 'mental blank' in dealing with some other subjects. Someone who is good at languages, for example, is often quite hopeless with mathematics, or mechanical subjects. Some people reason best with words, others with 'pictures' or diagrams.

It is just the same with logic. Some people will find one type of logic easy to understand and apply, but find other types of logic impossible to comprehend. So the answer to this particular problem is to work with the type of logic you understand best. At the same time, though, it does not necessarily follow that every type of logic can be applied to solving *any* problem in logic. Often the reverse is true. Solving a particular problem may involve a particular type of logic being used. Only mathematical logic, for instance, will give *exact* answers to purely mathematical problems.

Let's see how alternative types of logic can be applied to solving the following problem:

> George Brown, Tom Green and Bill White were talking. 'Funny thing', said Brown. 'I've just noticed we're wearing different coloured hats the same as our names.' 'Yes', said one of the others. 'But none of us is wearing a hat the same colour as our name. For instance, I'm wearing a green hat.'

Solution by Logical Reasoning
First write down the known facts or *premises*.

(i)   Brown, Green and White are each wearing a coloured hat.
(ii) The colours of the hats are brown, green and white.
(iii) Brown pointed out this fact.
(iv) The one wearing the green hat then pointed out that none of the colours of the hats they were wearing was the same as their names.

All these premises are *true*.

Analysis or argument: Since none is wearing the same colour as their name:

(a) Brown is wearing either a green or white hat
(b) Green is wearing a brown or white hat
(c) White is wearing either a brown or green hat
(d) Either Brown or White spoke last, and is wearing a green hat.

Conclusion: Since Brown had already spoken, it must have been White who spoke last (consistent with (d)). That means it was *White* who was wearing a *green* hat (consistent with (c)).

That leaves *Brown* wearing a *white* hat (consistent with (a) since the green hat has already been allocated). That leaves *Green* wearing the *brown* hat (consistent with (b) since the white hat has already been allocated).

> So – Brown was wearing the *white* hat
>   Green was wearing the *brown* hat
>   White was wearing the *green* hat

Now the same problem can be solved by several other different types of logic. If the following alternative solutions are not clear, read the appropriate chapter describing that type of logic first.

Solution by Simplified Boolean Algebra

Here we can allocate capital letters for the names – B for Brown, G for Green and W for White; and lower case letters for the colours of the hats – b for brown hat, g for green hat and w for white hat.

Let's concentrate on Brown, as the first name in the list. There is no need to write out the argument in full to mull over; we can express the relevant facts in simple equation form, noting that a $^-$ over a letter means 'not'.

$B = b$ or g or w (Brown is wearing a brown, green or white hat)
$B = \bar{b}$ (Brown is *not* wearing a hat the same colour as his name)
$B = \bar{g}$ (Brown is *not* wearing a green hat since the last person who spoke is White)

The $\bar{b}$ cancels the b and the $\bar{g}$ cancels out the g, so

$B = w$ (Brown is wearing the white hat)

The rest of the answer then follows in an obvious manner. The main point is that the mathematical solution is shorter, quicker, and positive.

Incidentally, a complete *proof* of this solution is given in the chapter on Boolean Algebra.

Solution by Logic Diagram

In this case the problem is set down in the following diagrammatic form:

	brown hat	green hat	white hat
BROWN			
GREEN			
WHITE			

Since the hat colours are not the same as the names we can fill in part of the diagram thus, using 'X' to show 'not':

	brown hat	green hat	white hat
BROWN	X		
GREEN		X	
WHITE			X

To proceed further we need another diagram to analyse the other available facts:

	spoke first	spoke last
BROWN		
GREEN		
WHITE		

Brown spoke first, so, that completes the first line with a √ and X (√ for 'yes' and X for 'not'). The person who spoke last was wearing a green hat. He cannot be Green, so he must be White. The second diagram can thus be filled in like this:

	spoke first	spoke last
BROWN	√	X
GREEN		
WHITE	X	√

(Green did not speak at all, so we could fill in his line with X and X, but this is not necessary for solving the problem.)

We can now go back to the original diagram and enter the fact that White has been identified as the last speaker, wearing the green hat:

	brown hat	green hat	white hat
BROWN	X	X	
GREEN		X	
WHITE	X	√	X

That leaves only one possibility for Brown. He must be wearing a white hat. The completed diagram also confirms that Green is wearing the brown hat:

	brown hat	green hat	white hat
BROWN	X	X	√
GREEN	√	X	X
WHITE	X	√	X

Solution by Block Logic

Here the possibilities are written down in the following form:

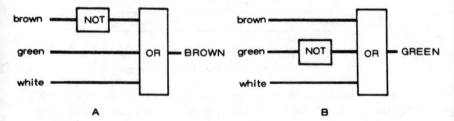

A   B

Fig. 2.1a,b

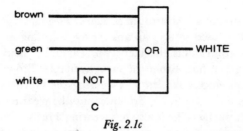

Fig. 2.1c

The person who spoke last was wearing a green hat, so this bit of information in block logic is:

spoke last ⟶ green 'Input' ⟶ BROWN **or** WHITE

But it cannot be Brown since he spoke first, which puts a NOT in the green 'input' line of diagram A:

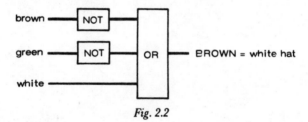

Fig. 2.2

The rest of the solution then follows by completing the diagrams for GREEN and WHITE (i.e. introducing a second NOT in the appropriate 'input' lines).

Incidentally, rendered in *Truth Table* form, as the complementary method of expressing block logic, the solution follows exactly the same lines as the solution by Logic Diagrams, using 1 and 0 instead of √ and X.

Solution by Arithmetical Logic

Not applicable since no numerical values are involved and there is no numerical answer.

Solution by Inductive Logic, Analogy or Heuristic Reasoning

Not necessary since *all* the facts necessary to obtain the solution are available.

Solution by Reductio ad Absurdum *or Indirect Proof*

This could be used to find an answer by assuming an answer and seeing if this contradicts the facts. For example, assume Brown is wearing the green hat. From the facts available, Brown spoke first. But the person who spoke *last* was wearing the green hat. Thus the assumption is not correct, i.e. Brown cannot be wearing a green hat. Equally, from the facts, he is also not wearing a brown hat. Therefore he must be wearing a white hat. The rest of the solution then follows.

This solution has been arrived at surprisingly simply. Note, however, it has used a *mixture* of logic. The final solution, starting with the conclusion that Brown must be wearing a white hat, is derived by deductive logic.

CHAPTER 3

Deductive Logic

In using the description *deductive logic*, the term deductive refers to the manner of reasoning or *argument*. It follows the same pattern as all approaches to logic in listing a number of *premises* (or statements in simple terms), which are then considered or 'argued' over to arrive at a logical solution to the problem or *conclusion*. Strange as it may seem at first, the 'logic' part is concerned only with the correctness or *validity* of *arguments* – not whether the premises themselves are true or false, or even whether the conclusion is true or false. But if the argument is valid (note: the argument is never itself said to be true or false), then the *conclusion* must be true if the *premises* are true. If the *argument* itself is logically incorrect or *invalid*, there is a possibility that the *premises* could be true and the *conclusion* false, or vice versa. Let's clarify this with some simple examples.

Example 1A
True premises and true conclusion

Premises:	All animals are mortal	(true)
	All cats are animals	(true)
Conclusion:	All cats are mortal	(true)

This is *valid argument*.

Example 1B

On the other hand we can have true premises leading to a true conclusion, but with *invalid argument*, e.g.

Premises:	All animals are mortal	(true)
	All cats are mortal	(true)
Conclusion:	All cats are animals	(true)

This is invalid argument, even if the conclusion is true, because there is nothing in the premises to establish the fact that cats *are* animals. In 'knowing' that they are, we are assuming another premise, not specifically given.

The *logical* correctness or incorrectness depends solely on the relation between premises given and conclusion. You cannot 'add in' another premise, however convenient. On the other hand, logical argument can ignore a false premise, as in the next example.

Example 2
True and false premises and true conclusion
Premises: All cats are mortal   (true)
                 All birds are cats   (false)
Conclusion: All birds are mortal   (true)

This is valid *argument* since its conclusion reached is true, even if one premise is false.

At this point it should be pointed out that although the examples quoted are simple statements this does not mean that deductive logic is restricted to simple, straightforward premises. Quite the reverse. The original subject may be quite complex and lengthy, the gist of which then has first to be extracted in the form of simpler premises. If dealing with more than one subject, it would also have to be broken down in order to provide conclusions for each subject separately. Simple statements are thus the end product of a preliminary assessment of the subject matter, as a preliminary to applying deductive logic. They express the gist of what is being analysed in the simplest possible terms, ignoring irrelevancies.

One thing needs to be made quite clear. Not every statement is necessarily a proposition or premise in logic. To qualify it must be capable of being either true or false. This can help explain some of the paradoxes in 'classic' logic, like the logic puzzle (antinomy) of Bertrand Russell's:

'In a certain town there is a barber who must shave all those people and only those people who do not shave themselves.' This leaves the question of whether the barber shaves himself or not. If he does, he breaks the rule and so he must not shave himself. If he does not shave himself, then he is again breaking the rule and so he must shave himself.

The answer to this is that the rule itself is wrong. It is both true *and* false in the sense that it both can and cannot be obeyed. (That can be arrived at by *deduction*.) There is another possible answer, perhaps an even better one. Logically, the barber cannot exist. (That can be arrived at by *intuitive logic*.)

Example 3
False premises and false conclusion
Premises: All cats have wings   (false)
                    All dogs are cats   (false)
Conclusion: All dogs have wings   (false)

This is still valid *argument* since the conclusion reached is a logical assessment of the facts presented in the premises, even if these are false.

Example 4
False premises are true conclusion
Premises: All cats have wings   (false)
                    All birds are cats   (false)
Conclusion: All birds have wings   (true)

Valid argument again!

At this point, deductive logic may appear to be anything but logical, but the above established the ground rules – the simplest and easiest to understand being: *If the premises are true, a valid argument must give a true conclusion.*

This then raises a most important point, that the *form* of the argument must also be valid for the argument itself to be valid. To explain this simply, let us take the premises of example 1A:

>   All animals are mortal
>   All cats are animals

Now cross-link the two statements:

>   All animals are mortal

>   All cats are animals

This reads as 'all cats are animals (and) all animals are mortal' which is a *valid form* of argument for concluding 'all cats are mortal'. In baisic 'formula' terms, this becomes:

>   All A are B

>   All C are A

(using A for 'animals'; B for 'mortal' and C for 'cats').

Now look at Example 1B, using letters again, which only cross-links like this:

All A are B
      ↑
All C are B

The only common feature is that both animals and cats are mortal. It is an *invalid form* to make a conclusion that all C are A (all cats are animals), even if the conclusion is correct. Substitute 'fish' instead of 'cats', for example:

All mammals are mortal
All fish are mortal

Conclusion: All fish are mammals.
The conclusion in this case is patently not true, because the form of the argument is invalid.

This, in fact, is one way of finding out whether the form of the argument is valid or invalid – see how they crosslink, and if 'substitution' to give an equally true premise (or a *counter-example*, as it is called), results in a conclusion which is obviously false.

This also explains why a true conclusion can be derived even when one or all of the premises are false – provided the *form* of the argument is valid.

Conditional premises
A conditional premise is in the form
    If A, then B

Which is then followed by a second premise A. The *valid* form of the argument is then cross-linked like this

If A, then B

A ———————↑

As a simple example:

Premises: If it rains, I will get wet
             It rains
Conclusion: I will get wet

Further *valid* forms are:

(i) If A, then B
     Not B
Conclusion: Not A
Example: *Premises:*   If it rains, then I will get wet
                          I have not got wet
          Conclusion: It is not raining

(ii) If A, then not-B
      B
Conclusion: Not-A
Example: *Premises:*   If it is fine, then I will not get wet when
                          I go out for a walk
                          I have gone out for a walk and got wet
          Conclusion: It is not fine

(iii) If not-A, then not-B
       Not-A
Conclusion: Not-B
Example: *Premises:*   If it is not raining, then I will not get wet
                          It is not raining
          Conclusion: I will not get wet

Invalid forms of argument are:

(iv) If A, then B
      B
Conclusion: A
Example: *Premises:*   If Bacon wrote Shakespeare, Bacon was
                          a great writer
                          Bacon was a great writer
          Conclusion: Bacon wrote Shakespeare.

(v) If A, then B
     Not-A
Conclusion: Not-B
Example: *Premises:*   If Bacon wrote Shakespeare, Bacon was
                          a great writer
                          Bacon did not write Shakespeare
          Conclusion: Bacon was not a great writer.

The validity, or otherwise, of premises and conclusion can also be proved mechanically, using Truth tables or Venn diagrams.

Categorical Statements

'Categorical' means unconditional, so a categorical statement is explicit, e.g. 'All A are B', or 'No C are D'. The former is an *affirmative* statement and the latter a *negative* statement, but both are explicit or *universal* as far as logic is concerned. But they are not necessarily expressing an unconditional *truth*. The statement 'all trespassers will be prosecuted', for example is categorical, but it does not necessarily mean that there *will* be trespassers – and the validity of being able to prosecute them if there were is something that does not necessarily hold true in law.

Other categorical statements in logic can be even more ambiguous or even contradictory if they are of the type:

Some A are B; or in negative form, Some A are not B

These are known as *particular statements,* thus proving that a categorical statement can be a true premise can be difficult at times. One way of doing this is to deny or 'invert' the statement, like this

statement: A
re-write as Not-A

Now if Not-A is shown to be true, the original statement A is false. On the other hand, if Not-A is false, then A is true. In other words, the statement itself is argued in valid logic. Equally, this is the basis of debate, where conclusion is finally decided by vote rather than truth!

Syllogisms

Syllogisms are arguments comprising only categorical statements and embracing two premises and one conclusion – the basic formula for deductive logic, in fact. Thus our original example 1A is a syllogism, for instance, and also the following:

Premises:	All cats are mammals
	All mammals are animals
Conclusion:	All cats are animals.

Now each of these categorical statements contains two *terms*, one of which is a subject and one of which is a predicate. One term occurs in each premise ('mammals' in the example) and is called a *middle term*. Each of the other two terms occurs once only in the two premises and once in the conclusion ('cats' and 'animals' in the example) are known as *end terms*. The rules of logic then state that for a syllogism to be *valid*:

(i) The middle term must be *distributed* exactly once
(ii) No end term may be *distributed* only once
(iii) The number of negative premises must equal the number of negative conclusions.

To apply, these, however, we have to understand clearly what 'distributed' means, and to do that we first have to be able to identify the subject and predicate in each categorical statement. Distribution then follows from whether the statement is universal or particular, affirmative or negative (refer back to the start of the section on 'Categorical Statements').

If the statement is *universal* and *affirmative*, then the subject is distributed and the predicate undistributed. If the statement is *particular* and *negative* then the subject is undistributed and the predicate distributed.

If the statement is *universal* and *negative*, then both the subject and predicate are distributed.

If the statement is *particular* and *negative* then the subject is undistributed and the predicate distributed.

Don't be confused at this stage. We can make everything much simpler (and more understandable-logic) by going back to our syllogism example:

All cats are mammals
All mammals are animals
All cats are animals

and re-write in simple 'formula' form using S for subject, P for predicate, and subscripts $_d$ and $_u$ to indicate distributed and undistributed terms, respectively. We also need to identify the *middle term* separately, which we will designate M.

All cats are mammals
becomes $S_d M_u$ (universal affirmative)
All mammals are animals
becomes $M_d P_u$ (universal affirmative)
All cats are animals
becomes $S_d M_u$ (universal affirmative)

Extract the 'formula' on its own

$$S_dM_u$$
$$M_dP_u$$
$$S_dP_u$$

(Now refer back to the definitions for the rules for a syllogism to be valid.)

Rule I is satisfied because the middle term is distributed exactly once (in the second line)

Rule II is satisfied because the first end term (S) is distributed twice; and the second term P is not distributed in either of its appearances. Thus no end term is distributed only *once*.

Rule III is satisfied because there are no negative premises and no negative conclusions.

The syllogism is therefore *valid*.

Now look at this syllogism:

    All men like football   (universal affirmative)
    No women are men   (universal negative)
Conclusion: No women like football (universal negative).

In 'formula' form this becomes:

$$M_dP_u$$
$$S_dM_d$$
$$S_dP_d$$

Rule I is broken because the middle term is distributed twice.

Rule II is broken because the end term P is distributed exactly once.

Rule III is satisfied since there is one negative premise and one negative conclusion.

The syllogism is thus *invalid*. (You could also deduce this from the fact that both premises are false, but that does not justify assuming that the conclusion is also false.)

Actually there is no need to proceed past the first rule in this case; any one rule broken means an *invalid* syllogism. In fact, if there are negative premise(s) or a negative conclusion, checking conformity to Rule III is the first thing to do. If this rule is broken the syllogism is invalid and there is no need to proceed further. Perhaps one more example to help your familiarity with syllogisms?

    All gemstones are valuable (universal affirmative)

Some diamonds are not valuable (particular negative)
Conclusion: Some diamonds are not gemstones (particular negative)

$$P_d M_u$$
$$S_u M_d$$
$$S_d P_d$$

Check Rule III first. There is one negative premise and one negative conclusion, so this rule is satisfied.

Now check the other rules.

Rule I is satisfied.

Rule II is satisfied.

The syllogism is therefore *valid*.

Deductive logic can, quite easily, introduce ambiguity. Here, for instance, is a classic example.

Ambiguity

A teacher of law made a contract with one of his pupils that the pupil would not have to pay for his lessons if he did not win his first case. The pupil completed the course of lessons, but did not take any cases. The teacher then sued for payment.

The pupil analysed his position logically, based on the following premises:

(i) Either I will win the case, or lose it.
(ii) If I win my case, I will not have to pay (the teacher will have lost his suit for payment).
(iii) If I lose my case, I will not have to pay since this is implicit in my contract with my teacher.

Since all three premises are true, the *logical conclusion* is that whether I win or lose my case I will not have to pay.

In a similar manner the teacher analysed his position, based on these premises:

(i) Either I will win the case, or lose it.
(ii) If I win my case, the pupil will have to pay me.
(iii) If I lose the case, the pupil will have to pay me under the terms of the contract because he will have won his first case.

Again, since all three premises are true, the *logical conclusion* is that whether I win or lose my case I will not have to pay.

Somewhere there is a false premise, leading to a false conclusion, in either the pupil's or the teacher's arguments. Or is there one in both? Or are both sets of premises and conclusions logically correct? It is a case of spotting the contradiction(s) or logical falsehood(s) – and finding the answer is not as simple as it seems. Perhaps, indeed, it is a subject for further argument and debate, with no logical conclusion at at all?

CHAPTER 4

Venn Diagrams

Venn diagrams – named after the nineteenth-century English logician and mathematician John Venn – are a pictorial method of representing, and analysing, deductive logic. They can be simple and effective to use – or confusing and difficult to understand. It depends, basically, on how you react to the principle involved.

In a Venn diagram each particular class (or what is also called a 'set') is represented by a fully enclosed shape. It does not matter what the actual shape is; it can be a circle, a square, a rectangle, or an irregular shape, provided it is a closed shape. For the purpose of explanation, elliptic shapes will be used for a start.

Any one shape then represents a class, e.g. in the first diagram of Fig. 4.1, this represents 'all A', or mathematically $A = 1$. If the shape

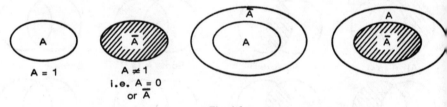

Fig. 4.1

is shaded, as in the second diagram, this excludes all A from being represented within this shape. Thus it represents $A \neq 1$, or $A = 0$, or $\bar{A}$ (not A). Enclose the shape within another shape as in the third diagram and all A is now represented by the smaller shape (and contained within it), so that everything outside and included within the larger shape is not A. Equally, if we shade the smaller shape to exclude A from it (making it $\bar{A}$), then all A is contained within the remaining plain area contained within the larger shape, as in the fourth diagram.

Fig. 4.2 extends this principle to two states, A and B, where the two shapes overlap. Thus in the first diagram this implies that some A are

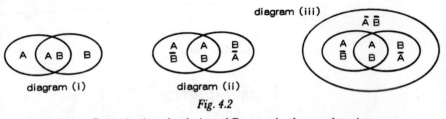

Fig. 4.2

B and some B are A, since both A and B occur in the overlapping area. At the same time A cannot occur outside its shape so is excluded from the non-overlapping part of the B shape. Complete labelling is then as shown in the second diagram. If these overlapping shapes were included within a larger shape, a further area would be produced from which both A and B are excluded, i.e. ĀB̄ within this area as in the third diagram.

Now look at the further diagrams shown in Fig. 4.3. These extend the principle of representing statements by Venn diagrams further. It is to be understood that the left hand shape designates A and the right hand shape B. By the rule established, A is excluded from the A shape

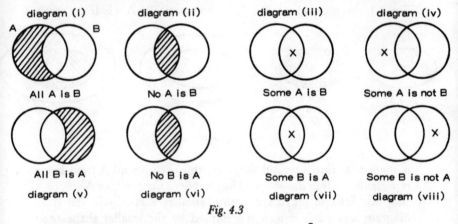

Fig. 4.3

by the shaded area in diagram (i) – which is then Ā – and only occurs in the overlapping area with shape B. Thus *all* the A is contained in this overlapping area, which is also included in shape B. Thus this diagram expresses the condition or premise '*all A is B*'.

Diagram (ii) excludes both A and B from the overlapping area (shaded). That means 'no A is B'; and equally, 'no B is A' – shown explicitly in diagram (vi).

Without shading, neither A nor B is excluded from the overlapping area, i.e. 'some A is B (diagram (iii)); and 'some B is A' (diagram (vii)). Since this is the *significant* part of the diagram this can be marked with an X.

Diagram (iv) shows a further relationship, with 'X' marking the significant part of the diagram – 'some A is not B' (some A lies outside the B shape). Similarly, diagram (viii) displays 'some B is not A'.

Blank parts of a Venn diagram mean nothing on their own until all the available information has been entered, either by shading (denoting exclusion from that particular area), or by an 'X' or 'some' relationship, or both. Thus exclusion is expressed by a *universal* statement, e.g. 'all A are ...'; and an 'X' designates a particular relationship, e.g. 'some A are ...'.

Venn Diagrams and Syllogisms

Venn diagrams using two overlapping shapes, as just discussed, are basically a diagrammatic representation of two categorical propositions. If a third overlapping shape is added the resulting Venn diagram can be used to test syllogisms, and/or solve problems in deductive logic. This time, to establish the overlapping areas neatly the shapes are best drawn as circles.

As a simple example, take the premises:

  (i) All animals are mortal
  (ii) All cats are animals

(In the chapter on Deductive Logic we have seen that the *conclusion* in this case is 'all cats are mortal'; and the argument is *valid*.)

Treating this in Venn diagram form we draw three overlapping

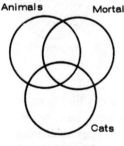

diagram (i)

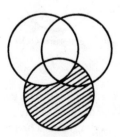

diagram (ii)

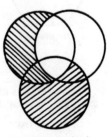

diagram (iii)

Fig. 4.4

circles, one for *animals*, one for *mortal* and one for *cats*, as in the first diagram of Fig. 4.4. Now, *all* cats being animals excludes cats from that part of the cat circle which is outside the *animals* circle. Shade that area in as in diagram (ii).

Equally, all animals being mortal excludes animals from the animal circle outside the mortal circle, so shade that area in, as in diagram (iii). We are left with the conclusion that the only area now accommodating *all cats* is within the *mortal* circle - hence 'all cats are mortal'.

Now try an example where the premises are true but the conclusion drawn (by deductive logic) is false:

Premises: (i) All dogs are mammals
              (ii) All cats are mammals
Conclusion: All cats are dogs.

This time shading-in the diagram produces the result shown in Fig. 4.5 (last diagram). The only area where *all* the cats are contained is within the mammal circle. Cats are excluded from the *dogs* circle by the shading. Therefore, from the Venn diagram, *no* cats are dogs. Testing the (deductive) conclusion in this way shows that the argument in this case is not valid.

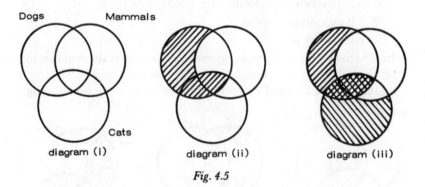

Fig. 4.5

Two premises imply *three* specific terms, each one of which is represented by a circle in a *Venn* diagram. The *conclusion* can be tested by, or deduced from, the final diagram. Thus a Venn diagram is essentially a diagram or premises or propositions embracing a subject (S) or major term; a predicate (P) or minor term; and a middle term (M). A similar rule applies as for deductive logic - for valid argument the middle term must be distributed exactly once, i.e. appear unshaded in each of the other terms (circles). Conventionally, too, the middle term (M circle) is placed at the bottom of the diagram.

The premises used in the previous examples have been *universal* premises, i.e. of the form *all* A are B. The case of a *particular* premise, i.e. *some* A are B needs a little more explanation. For this, take the following which consists of one universal premise and one particular premise:

  (i) *All* artists are gifted people
 (ii) *Some* artists are poor
Conclusion: Some poor people are gifted people.

The Venn diagram for this is developed as shown in Fig. 4.6. The first (universal) premise excludes all artists from the artist circle outside the gifted people circle, so this part of the artist circle is shaded. Only *some* artists are poor, though, so this area of overlap is marked with an X. This 'some' area is common to artists, poor people and gifted people. Hence some people can be artists (as given by the premise); and equally some poor people can be gifted (as given by the diagram).

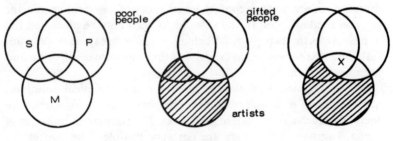

Fig. 4.6

In other cases the allocation of the 'X' area for 'some' may not be clear from the premises, which will indicate that the argument is *invalid* (although the conclusion reached by deductive logic may or may not be correct). For this futher example, consider:

Premises: (i) all good mathematicians are good scholars
              (ii) some athletes are good scholars
Conclusion: Some athletes are good mathematicians.

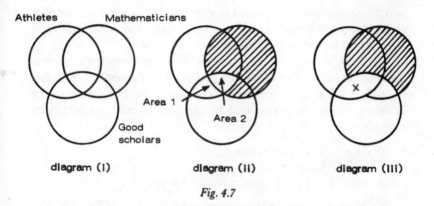

Fig. 4.7

In the corresponding Venn diagram – Fig. 4.7 – mathematicians are excluded from their circle except where it overlaps the good scholars circle by virtue of the universal premise (i). *Some* athletes are good scholars (from the particular premise), but there are two overlapping areas where this could be marked with 'X', as in diagram (ii).

There is not enough information in the premise to indicate whether the 'X' should be in area 1 or area 2. Area 1 seems the logical placement, for it *is* known that some athletes are good scholars. But there is no information to confirm whether or not *some* mathematicians are also good athletes. In this case the 'X' can only be placed on the line between the two (it could belong to either area), as in diagram (iii). The Venn diagram, in fact, shows that this form of syllogism and argument is invalid.

Venn diagrams can also be used for mathematical solutions, although there are simpler and more effective diagrams, or alternative methods which can be used in such cases. The particular limitation of Venn diagrams is that they are not very flexible. The greater the number of terms to be accommodated the more unwieldy (and possibly confusing) the diagram becomes. This virtually limits their application to a maximum of four separate terms or variables (i.e. four overlapping shapes).

Lewis Carroll Diagrams

Lewis Carroll – an excellent mathematician as well as author – developed an alternative form of logic diagram using squares. Fig. 4.8 shows a Carroll diagram plotted for two variables and two conditions

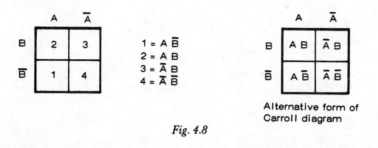

Fig. 4.8

for each (A and not-A, B and not-B). The diagram then plots in full the various possible combinations – four in this case.

Unlike a Venn diagram, this type of diagram can be extended to accommodate quite a number of separate terms before it becomes excessively complicated. It is, in fact, basically the same as a Karnaugh map – *see* Chapter 13.

CHAPTER 5

Simple Logic Diagrams

Many people find it easier to solve logic problems by diagrams rather than deductive reasoning or mathematical solutions – 'pictures' speak louder than words. Basically, however, a logic diagram is a Truth Table, complementary to block logic and Boolean algebra. But for the non-mathematically minded person such a diagram can be divorced from these more obscure subjects by plotting the diagrams on a 'yes' or 'no' basis, using √ for 'yes' and X for 'no' instead of the more formal symbols 1 and 0, respectively. A large proportion of problems designed specifically as 'logic puzzles' can be solved by using diagrams of this type.

Here is a simple puzzle-problem which can be solved using a logic diagram:

Smith, Jones and Thomas live in London, Brighton and York, not necessarily in that order. All travel away from home to work.

Smith travels to London to work.

Thomas lives further South than Smith.

Jones also works in London.

Find out where each lives.

The basic logic diagram is drawn up like this:

	London	Brighton	York
SMITH			
JONES			
THOMAS			

Smith travels to London to work, so he does not live in London. Put an X against Smith under London.

	London	Brighton	York
SMITH	X		
JONES			
THOMAS			

Jones also works in London, so does not live in London. Put an X against Jones under London. The remaining space in this column is then filled with a tick – the only remaining possibility, i.e. Thomas lives in London.

	London	Brighton	York
SMITH	X		
JONES	X		
THOMAS	√		

Now Thomas lives further South than Smith. Since Thomas lives in London that means Smith must live in York. The York column can now be completed. The Xs in this column merely confirm that Thomas does not live in York (he lives in London); and since Smith lives in York, Jones cannot.

	London	Brighton	York
SMITH	X		√
JONES•	X		X
THOMAS	√		X

The only possibility left is that Jones lives in Brighton, so the diagram is completed like this

	London	Brighton	York
SMITH	X	X	√
JONES	X	√	X
THOMAS	√	X	X

Thus: Smith lives in York
        Jones lives in Brighton
        Thomas lives in London.

Note the *principle* of working with such diagrams. Once a √ (yes) has been established in any horizontal line or column, then the *remaining* spaces in that particular line or column can be filled with Xs. Similarly, if all the spaces but one in a line or column are filled with Xs, then the remaining space must be a √ (yes).

More Facts – More Diagrams

Sometimes the facts available need treating in separate diagrams to solve alternative or complementary possibilities. Here is a further example illustrating this. Problem: There are three people with the names Smith, Jones and Robinson. Their Christian names are Arthur, Mary and Jane, not necessarily in that order.

  (i) Their ages are 17, 24 and 30.
 (ii) Miss Jones is 7 years older than Jane.
(iii) The person named Smith is 30 years old.

What are their full names and ages?

(Note: after the original statement, the facts have been separated for ease of reference. In a set problem the facts may be incorporated in one complete statement. It is then first necessary to separate them out as individual facts.)

To solve this particular problem, draw up diagram A relating surnames to Christian names. The only immediate clue is that *Miss* Jones will not have the Christian name Arthur, so an X can be marked in the Arthur column against Jones.

	Arthur	Mary	Jane
SMITH			
JONES	X		
ROBINSON			

(diagram A)

There seems to be more information available about ages, so construct two more diagrams on this basis:

	17	24	30
SMITH			
JONES			
ROBINSON			

(diagram B)

	17	24	30
Arthur			
Mary			
Jane			

(diagram C)

Now enter the main facts:

(a) Smith is 30 years old – enter in diagram B and complete SMITH line.

(b) Miss Jones is 7 years older than Jane. The only possibility is that Miss Jones is 24 and Jane is 17.

Complete JONES line in diagram B and the Jane line in diagram C.

	17	24	30
SMITH	X	X	✓
JONES	X	✓	X
ROBINSON			

(diagram B)

	17	24	30
Arthur			
Mary			
Jane	✓	X	X

(diagram C)

Diagram B can now be completed:

	17	24	30
SMITH	X	X	✓
JONES	X	✓	X
ROBINSON	✓	X	X

Diagram C can also be completed by filling in the top line with the only possible alternative (Arthur is 30)

	17	24	30
Arthur	X	X	✓
Mary	X	✓	X
Jane	✓	X	X

From these completed diagrams we can now complete diagram A

	Arthur	Mary	Jane
SMITH	✓	X	X
JONES	X	✓	X
ROBINSON	X	X	✓

Answer to problem: Arthur Smith is 30
                   Mary Jones is 24
                   Jane Robinson is 17

Instead of separate diagrams the whole problem can be entered on one diagram, appending the additional diagram(s) required to the right and bottom of the main diagram. In this case the logic diagram would be:

	Arthur	Mary	Jane	17	24	30
SMITH						
JONES						
ROBINSON						
17			
24			
30			

This would fill in, step-by-step, in the same manner, transferring information from the side and bottom diagrams to the main diagram as it becomes available.

CHAPTER 6

Inductive Logic

Inductive logic is similar to deductive logic in that conclusions are based on premises, but with one very important difference. The conclusions extend beyond the area of fact provided by the premises. Similar rules apply as to what is valid argument and a true conclusion, but the conclusion is necessarily qualified. It is not *necessarily* absolutely correct, particularly as it often has to be based on incomplete facts. It may be a convenient generalization, or even a forecast by extrapolation.

The 'safest' type of inductive arguments providing conclusions (i.e. those least likely to be in error) are those based on observed data, but even these have numerous pitfalls, correctly called *fallacies*. Take as an example a spring suspended vertically and carrying a pan into which weights can be put. The downward deflection of the spring is measured with weights of 10, 20, 30, 40, 50 grams, etc., added to the pan and plotted as points of a graph – Fig. 6.1. These points all lie in a line, so inductive argument would conclude that the relationship between deflection and load is a straight line, and the graph drawn in accordingly. It would also seem logical that if this relationship is true it will

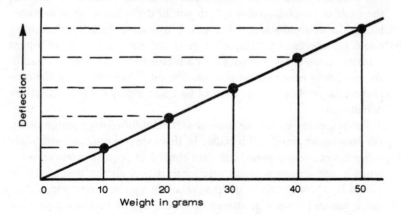

Fig. 6.1

extend beyond the measured range, i.e. downwards to give zero deflection with no load; and upwards to forecast the deflection produced by larger loads.

In fact, this conclusion holds true, up to a point. It could be used to calibrate a spring and a pan as a simple weighing machine, but there are other factors (not catered for in the original premise) which could modify the truth of the conclusion – Fig. 6.2. There could be a physical

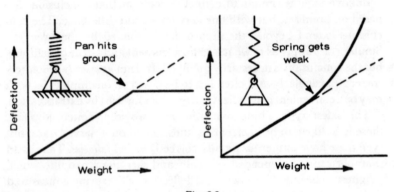

Fig. 6.2

limit to the amount of spring deflection before the pan hits something, for instance. Or if there were no physical limit to movement, the spring could be stretched beyond its normal behaviour pattern (i.e. exceeds the limit of proportionality). Both would deny the truth of extrapolating the original straight line graph upwards. Equally, the spring could be affected by 'fatigue' in repeated use, so that it no longer returns to zero deflection with no load. However, its behaviour within the originally observed range is unaffected. Thus extending the conclusion *too far* beyond the area of fact has nullified the truth of the conclusion.

In fact, even within the area of fact established by measurement, the conclusion could still be false. In this case it is not, but with some other system where measured data spaced at regular intervals show what is apparently a linear relationship (i.e. all the points can be joined by a straight line), the true relationship could well vary differently *between* points, e.g. instead of a straight line between points the true representation could be a *curve* waving up and down.

This point is made more clearly in Fig. 6.3, which shows a simple wave form. Say this represents an alternating current, the value of which is read by a meter. Suppose measurements are made at intervals 1, 2, 3, etc., which happen to correspond in time exactly to the peaks of positive current. This would give points lying in a straight line apparently indicating a steady positive current, Fig. 6.3 diagram (A). Equally, if they were made with the same interval but starting at a quarter of an interval later (1', 2', 3', etc.), all the measurement points would indicate zero current Fig. 6.3 diagram (B). Started half an interval later, they would indicate a steady negative current.

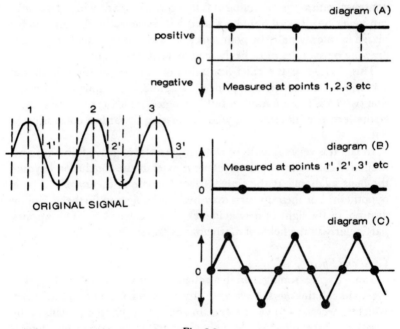

Fig. 6.3

This is an error (or what would be called in inductive logic terms a *fallacy*) due to unrepresentative *sampling*. Not enough sample measurements have been taken to give anything like a true picture, so the inductively reasoned conclusion is false. Take measurements at shorter intervals and this will immediately show up. Taken at one quarter of the original time interval, for example, the measured points and the subsequently induced plot would zig-zag up and down – Fig. 6.3 diagram (C). Still not true, but a much closer representation.

It is lack of sufficient samples (data), or *unrepresentative* samples which is a common reason for false conclusions in inductive logic, particularly when applied to statistics. Opinion polls are a classic example. They are based on answers (data) provided by a limited number of people, yielding, say, 65 per cent in favour of something and 30 per cent against (the other 5 per cent 'don't know'). The pollsters conclude, by induction, that 65 per cent of the *whole population* are in favour, 30 per cent against. As a matter of *logic*, this is basically nonsense, unless the sampling is fully representative (which in fact it can never be). It can only be a *statistical generalization*. It can also be biased unintentionally by poor selection of samples; or deliberately by drawing on samples most likely to give the *desired* result.

This, indeed, is the chief limitation of inductive logic – basing a conclusion on insufficient evidence (data) and/or jumping to a conclusion (which is human nature, but not logic). At least these are honest errors (errors of ignorance), whereas biasing towards a conclusion is not.

In fact it is very difficult, or even impossible, to prove fully that any inductive conclusion is true because it extends the conclusion beyond the facts available at the time. It can become a 'theory of ...' or a hypothesis; but these in turn may become subject to further proof or disproof in the light of new evidence, *or a new hypothesis*. This applies particularly in the fields of mathematics and science.

Cause and Effect

Inductive reasoning can be called a study of *cause* and *effect* – logicians call this *causal connections*. It is based on a condition (or cause) which is necessary to yield a certain effect, leading to a conclusion or *hypothesis* relating the two. As such it is an inductive *generalization* which may or may not be completely true.

Taking a simple everyday example, it is well known that paper or wood will burn in air, which contains oxygen. But it will not burn in a sample of air from which the oxygen has been removed. The inductive generalization which could be drawn is that oxygen (cause) produces combustion (effect). However, this is only part of the true picture. To burn in the presence of oxygen the paper or wood must first be raised to a suitable temperature, e.g. by holding a match to it.

To be more complete the generalization would need to take in this second cause (temperature) into the final statement.

A working 'formula' for such a method of reasoning would be – cause 1 (oxygen) in the presence of cause 2 (a suitable temperature) produced combustion of paper and wood (effect).

The other thing to emerge is that causal connections are based on a study of observed 'happenings' – what happens (i.e. what is the effect) in a particular circumstance or condition (cause). Equally important is how many times this occurs. The more times the greater the likelihood of the (generalized) conclusion being true, or at least substantially true. This can be expressed in the following form:

Instance 1 of happening A is related to cause B present
Instance 2 of happening A is related to cause B present
Instance 3 of happening A is related to cause B present
. . . .
. . . .

The conclusion to be drawn from this is:
All instances of happening A are accompanied by cause B.

Note: This conclusion is reached partly by inductive reasoning and partly by analogy, induction and analogy being closely related.

Causal connections summarized in this way can at best be valuable, and at least be merely suggestive. They can also lead to conclusions which are not *factually* correct, but despite this may be acceptable to quite a number of people. Here is a classic example, using the form given above:

Arthur broke a mirror, cut his hand, which was bad luck.
Mary broke a mirror, lost her purse, which was bad luck.
Tom broke a mirror, sprained his wrist, which was bad luck.
Conclusion drawn: breaking a mirror *causes* bad luck.

Testing Causal Connections
There are, however, methods of testing causal connections and resulting conclusions, known as Mill's 'canons' (or rules), after the English philosopher of that name.

(i) Method of Agreement
In words: if two or more instances of a happening under investigation

have only one circumstance in common, the circumstance in which alone all the instances agree is the cause (or effect) of the given happening.

Represented in symbol form, using capital letters for circumstances and lower case letters for happenings, this becomes:

A B C D occur together with a b c d
A E F G occur together with a e f g
Therefore A is the cause (or effect) of a.

(*ii*) *Method of difference*

In words: if an instance of a happening occurring and an instance in which it does not occur have every circumstance in common but one, that one occurring only in the former; the circumstance in which alone the two instances differ is the effect (or cause) or an indispensable part of the cause of the happening.

In symbol form:

A B C D occur together with a b c d
B C D occur together with b c d
Therefore, A is the cause (or effect), or an indispensable part of the cause of a.

(*iii*) *Joint Agreement and Difference*

This combines both the Method of Agreement and Method of Difference and in symbol form becomes:

A B C ----- a b c   A B C ----- a b c
A D E ----- a d e   B C ----- b c
Therefore, A is the effect (or cause), or an indispensable part of the cause of a.

(*iv*) *Method of Residues*

Rather cumbersome in words, this is: subtract from any happening such part as is known by previous induction to be the effect of certain antecedents, when the residue of the happenings is the effect of the remaining antecedents.

In symbols this becomes much simpler:

A B C --------------- a b c
B is known to be the cause of b
C is known to be the cause of c
Therefore, A is the cause of a.

(v) *Method of Concomitant Variation*

In simple language, this accommodates varying happenings designated by + or − signs and is easiest to understand in symbol form:

A B C   --------------- a b c
A +B C --------------- a ±b c
A −B C --------------- a ±b c

Therefore, A and a are causally connected.

Although looking less understandable at first sight, this is actually the most useful of the methods of testing causal connections since it can provide *quantitative* analysis of inductive reasoning.

CHAPTER 7

Simple Arithmetic or Logic with Numbers

Many problems involving numbers or quantities, can be solved by simple 'schoolboy' maths. The first rule is that to find two unknown quantities there must be two separate statements relating those quantities.

Example: Arthur is twice as old as Bertie. If their combined ages is 54 years, how old are they?

There are two statements, so the problem is solvable.

Write down the two statements in simple equation form, using A for Arthur's age and B for Bertie's age:

(i) $A = 2B$
(ii) $A + B = 54$

Equation (i) gives the *equivalence* between A and B directly, so use this fact to rewrite equation (ii):

$$A + B = (2B) + B = 54$$
$$\text{or } 3B = 54$$
$$\text{Hence } B = 18 \text{ years}$$

And since $A = 2B$

$$A = 2 \times 18 = 36 \text{ years}$$

This is a very elementary example but it is surprising how often the *equivalence* can be extracted from one equation (statement) and directly substituted in the other equation (statement) to obtain the answer to one of the unknowns. The answer to the other unknown then follows quite simply, using either of the equations.

The method is readily recognizable as solutions to miscellaneous equations – a subject that clicked or not at school. But what happens when more than two unknowns are involved? Does this simple mathematical method work? Not necessarily, unless you apply a little logic, as the following will show:

There are a mixture of red, green and blue balls in a box. The total number of balls is 60. There are four times as many red balls as green balls; and 6 more blue balls than green balls.

We have three different statements, so we can write three different equations, using R for red, G for Green and B for blue; and x, y and z for the *number* of red, green and blue balls respectively. (A mathematician would probably set about it that way.)

Equation (i): (x times R) + (y times G) + (z times B) = 60
or using a . sign for multiplication:

$$x.R + y.G + z.B = 60$$

There are 4 times as many red balls as green balls, so

Equation (ii): $x = 4y$
There are 6 more blue balls than green balls, so

Equation (iii): $z = y + 6$

we now have three equations to manipulate to solve for three unknowns. They are simple equations in this case and easy to solve by substitution, but the problem has got a little confused by introducing x, y and z. So let's start again with a logical approach.

There are more blue balls than green balls (one fact)
There are more red balls than blue balls (another fact)
That means that the green balls must be the least in number (conclusion).

Now let's put this together in simple arithmetical form, forgetting about the x, y and z approach, and using R, B and G for the *numbers* of red, blue and green balls, respectively. Use the number of green balls (G) as a base, because they are the least number

number of green balls = G
number of red balls = 4 times as many green = 4G
number of blue balls = number of green balls + 6 = G + 6

Now together $G + 4G + (G + 6) = 60$

$$\text{or } 6G = 60 - 6$$
$$= 54$$
$$\text{hence } G = 9$$

It then immediately follows that

$$R = 4G = 4 \times 9 = 36$$
$$B = G + 6 = 9 + 6 = 15$$

The point of this example is not to over-complicate the problem by writing out comprehensive mathematical equations. Approach the problem logically, looking for the simplest solution. The original mathematical equation given is, in fact, a redundant statement. It introduces x, y and z for unknowns, and at the same time retains R, B and G. (In fact x = R, y = G and z = B.)

This is a basic error mathematicians often make – over-complicating translation of facts into equations so that redundancies are introduced, or the whole equation becomes too complex for easy solution. At the extreme, mathematical analysis can become so complex that only the originator can understand the working – and as a result nobody can check the result!

Suppose, now, the information given in the problem is incomplete, i.e. using the balls in a box example again, only two facts are given, the total of 60 balls and the number of red balls being four times the number of green balls. There is no mathematical solution in this case. However, logic – and a little simple arithmetic – can give us the various numbers of different coloured balls *possible* within the facts.

(i)   The relationship between red balls and green balls is known ($R = 4G$). Also there are three different colours of balls in the box, so there must be at *least* 1 blue ball.

(ii) To be consistent with the fact that $R = 4G$ means that the total of R and G together ($R + G$), must be a multiple of 5. (4 parts for R plus 1 part for G.)

(iii) The largest *available* total is $60 - 1 = 59$ (there must be at least 1 blue ball). This is not divisible by 5. So the largest *possible* total ($R + G$) is 55.

(iv) That means there must be *at least* 5 blue balls, leaving 55 to be allocated between R and G on a 4 to 1 basis, i.e. $R = 4 \times 11 = 44$ and $G = 11$.

That is our first possible solution: $B = 5$
$$R = 44$$
$$G = 11$$

(v)   The *next* highest possible total is 50 (again divisible by 5). This gives a second possible solution:

$$B = 10$$
$$R = (4 \times 10) = 40$$
$$G = (1 \times 10) = 10$$

(vi) By the same reasoning, the next highest possible total is 45, which gives the following answers:

$$B = 15$$
$$R = (4 \times 9) = 36$$
$$G = (1 \times 9) = 9$$

and so on, until we come to the last *possible* answer when there are 55 blue balls:

$$B = 55$$
$$R = (4 \times 9) = 4$$
$$G = (1 \times 1) = 1$$

There is no *correct* answer produced by this form of logic. There are not enough facts to establish one. In other words, on the information available, it is only possible to derive all *possible* answers.

This method of deriving *possible* solutions by logic will also work when there are enough facts available for an exact solution, using only some of the facts. The exact solution is then obtained by seeing which of the possible solutions fits the remaining fact(s). This is a more lengthy process, of course, but in certain circumstances can be a simpler one when it seems difficult to assemble *all* the available facts in equation form.

Real life problems, too, may not contain all the facts for exact solution. In that case, only *possible* solutions can be obtained. Numerical problems set as puzzles or 'mind-teasers' on the other hand invariably contain all the facts for exact solution, although this does not necessarily preclude the possibility of there being alternative possibilities.

Illogical Logic

Consider the following: Johnson said 'If my son's age was trebled he would be as old as me. But in 15 years time he will be half my age'. How old are we both now?'

First let's solve this by logical arithmetic, using S for son's age and J for Johnson's age. Assembling the equations from the facts:

Equation (i) $3 \times S = J$
Equation (ii) $S + 15 = \frac{1}{2}(J + 15)$

to make it simple, re-write this as

Now since $J = 3 \times S$ from equation (i), substituting this in equation (ii) gives

$$2 \times S + 30 = 3 \times S + 15$$
or $$15 = S$$

Thus if the son's age is 15, Johnson must be 45.

Suppose now the problem was a little more obscure, or we could not see how to derive the necessary two equations. In that case try solving the problem on an arithmetical trial-and-error basis, using a little genuine logic and a lot of 'illogical logic'.

If Johnson is talking about 15 years on, he is probably now under 60. His son, too, will probably be 20–30 years younger than Johnson. On this basis, let's *assume* that Johnson is 54, when his son's age must be 18.

What does this give in 15 years time?

Johnson will be 69 and his son 30. This does not agree with the second fact – it shows a surplus of 9 years. So *logically* Johnson must be younger than 54 – and *illogically* by that surplus. Thus Johnson is $54 - 9 = 45$, and surprisingly that checks out as correct!

Do not expect 'illogical logic' to provide correct answers every time. It certainly won't. But it is the basis of another method of handling arithmetical logic. First an answer is estimated (or guesstimated, if you like) on a reasonably logical basis, i.e. what appears to be the right *order* of answer. Then see how it fits the facts. From this, adjust the original estimate up or down to get nearer to the facts, until eventually your adjusted answer meets the facts.

What you are doing, indeed, is employing a sort of loose variation of *reductio ad absurdum* together with *logical reasoning*.

Not to be recommended for general use – 'illogical logic' can become tedious if dealing with two or more unknowns which have to be estimated and continually readjusted. But when all else fails and you cannot see how to work out the arithmetical problem, you could try it instead of giving up entirely!

Devise Your Own Formulas

Quite often at work or business similar tedious calculations crop up at various times, which have to be worked out from first principles. A desk calculator helps, but it still takes time. How very much simpler if a working formula can be devised to cope with the problem.

As a hypothetical example – and an exercise in logic – let's take the case of a printer who produces large catalogues for various clients. Trying to cut costs, individual clients often ask how much can they save by leaving off the page numbers? The printer's cost for setting these numbers is, say £5 per thousand digits. A particular client asks what the savings would be on his 512-page catalogue.

The printer now sits down to think, and works things out like this:

  (i) Pages 1 to 9 require one digit for page numbering, making *9 digits*.

 (ii) Pages from 10 to 99 require 2 digits for each number and thus 20 digits for each ten pages, requiring $9 \times 20 = 180$ *digits*

(iii) Pages from 100–999 require 3 digits for each number and thus 300 digits for each hundred pages. Four groups of hundred pages will take the book up to 499 pages, requiring $4 \times 300 = 1200$ *digits*

(iv) Pages 500–512 will then require 3 digits per page, on 13 pages or $3 \times 13 = 39$ *digits*.

Adding all this up he finds 1428 digits required, which at a cost of £5 per thousand amounts to a sum of £7.14 if the page numbers are omitted. (His time in working it out will have been worth more than that!)

Logically, when faced with a similar problem again he will simply say to the client (from experience) that any saving is negligible. But suppose the client persists in asking what the saving is for a particular book. Does he go through the whole process again? Not if he has spotted the *formula* which can be derived from the original working. The number of digits required fall into a distinct pattern:

$$\text{pages } 1\text{–}9 = 1 \times 9$$
$$\text{pages } 10\text{–}99 = 2 \times 90$$
$$\text{pages } 100\text{–}999 = 3 \times 900$$

and so on (e.g. pages 1000–9999 for a really long book used $= 4 \times 9000$ digits). So the basic formula is:

number of digits =	(1×9)	$+ (2 \times 90)$	$+ (3 \times 900)$	$+ (4 \times 9000)$
pages covered	9	99	999	9999

Any *intermediate* number of pages can then be inserted into this formula for a quick answer.

Lets use 512 pages again, which falls in the third group. The fourth group does not apply and is ignored. Put the actual total of pages in their third group, and deduct the prior pages covered by the previous two groups:

$$\text{number of digits} = (1 \times 9) + (2 \times 90) + 3 \times (512 - 99)$$
$$= 9 + 180 + 1239$$
$$= 1428$$

Is it Possible?

Here is another example of a problem in arithmetical logic where it is not known at the start whether there *is* a possible answer or not.

Bellamy is travelling to America. He has bought twenty-five $100 notes from his bank. For safety he wants to distribute them in different pockets, with each pocket holding a different number of notes and different total value. He has seven pockets in all. Can he do what he wants?

This problem is tackled from first principles. The *least* number of notes he can have in any one pocket is 1; and the *least difference* in the amounts in each pocket must also be 1 (in this case 1 note). So write down numbers (i), (ii), (iii), etc, representing pockets and enter under each the number of notes allocated on the above basis.

pockets	(i)	(ii)	(iii)	(iv)	(v)	(vi)	(vii)
notes	1	2	3	4	5	6	7

Add up the number of notes needed to do this. The total is 28 notes. But Bellamy only has 25 notes, so he cannot fulfil his original plan.

How, in fact, could he do it, for he cannot afford to buy more dollars. Only changing one (or more) $100 notes into smaller denominations so that he can have a greater number of notes.

The answer is quite simple in this case. Logical assessment of the position has established that he needs 28 notes, not 25. So Bellamy changes three $100 notes into six $50 notes. That now gives him a total of 28 notes, which he can now allocate like this:

pockets	(i)	(ii)	(iii)	(iv)	(v)	(vi)	(vii)
number of notes	1	2	3	4	5	6	7
value of notes	$50	$50	$50	$100	$100	$100	$100
money in pocket	$50	$100	$150	$400	$500	$600	$700

This could have been tackled in another way. The basic 'plan' for allocating notes into different pockets shows that a total of 28 notes are required. So it was rather foolish of Bellamy to have spent all his money in buying $100 notes for with the amount he had available he could only get 25 notes. He should have looked for the answer first, and saved himself the trouble of going back and changing three notes. He was also fortunate that there *was* a simple solution to his requirements.

Jackson was much more methodical about planning his requirements for his trip to America. He hit on the same basic plan, and had the same amount of money for buying dollar notes, but he went about working it out this way.

(i) The *least* amount of notes I need are 28; and I can afford to buy up to $2500 worth of notes.

(ii) I cannot afford to buy twenty-eight $100 notes as that totals $2800 – more than I have.

(iii) I will therefore buy twenty-eight $50 notes which will cost me $1400 and leave me $1100 still available

(iv) I will then allocate these notes: 1 in pocket (i), 2 in pocket (ii), etc.

(v) I can afford to buy eleven more $100 notes. If I add one to each pocket, that will still give me a different number of notes and different total value in each pocket. That will need seven more notes, costing $700 and leaving me with $400 unspent.

(vi) If I use this to buy four more $100 notes I can still conform to my original plan by putting all four of these additional notes into either pocket (iv), (v), (vi) or (vii); or alternatively one extra $100 not into each of these four pockets.

Jackson has arrived at possible solutions by a mixture of simple arithmetic and deductive logic.

Quite often problems of this type do *not* have an answer – *see* the example under *Reductio ad absurdum* (Chapter 1).

Looking for Short Cuts

Here is what looks like a tedious example of solving simultaneous equations for unknowns A, B, C and D. A solution *is* possible (assuming all the equations are valid) because there are four unknown and four separate equations.

(i) $A + 2B + 3C + 4D = 2$
(ii) $4A + 4B + 2C + 2D = -2$
(iii) $5A + 2B + C - 3D = -6$
(iv) $3A - 2B - C + 7D = 22$

Let's look for a short cut. There is one obvious one we can try. Adding equations (i) and (ii) together gives:

$$5A + 6B + 5C + 6D = 2 - 2 = 0$$

Now group as:

$$5(A + C) + 6(B + D) = 0$$

Thus $A + C = 0$, which means $A = -C$
                              or $C = -A$
and $B + D = 0$, which means $B = -D$
                              or $D = -B$

Try substituting for C and D in equations (iii) and (iv) which then become:

(iii) $5A + 2B - A + 3B = -6$ which simplifies to:

$$4A + 5B = -6$$

(iv) $3A - 2B + A - 7B = 22$ which simplifies to:

$$4A - 9B = 22$$

We now have two simple simultaneous equations to deal with. Subtracting (iv) from (iii) gives:

$14B = -28$
or $B = -2$, when D must equal 2

Substitute $B = -2$ in another equation, say (iii):

$4A - 10 = -6$
      so $A = 1$, when C must equal -1

Smarten up your school algebra, if it has got a bit rusty. It can be the quickest and simplest method of solving problems or puzzles in logic involving numbers.

CHAPTER 8

Logic in Aptitude Tests

Aptitude tests are widely used in vocational guidance and industrial training. To be *valid* aptitude tests the questions must be solvable by a specific *talent*, not by knowledge or learning. Some, but only some, can be solved by *logic* reasoning. Others have to be tackled by mental reasoning, numerical reasoning or abstract reasoning. Still more are based on latent technological skills (fitting pegs into different hole shapes as an elementary example), but these are outside the scope of a book on logic.

An example of test questions which can be solved by logical reasoning is diagrams of a series of different shapes in different sizes and colours arranged in a pattern, but with some shapes missing. The problem is to find what the missing shapes should be. Since the complete pattern is planned (by the designer of the test) on some logical basis, there must be a logical answer.

The following is a very elementary test; the question to be solved being what are the next three symbols in the sequence:

□ – ○ – △ – □ – ○ – △ – □ – ○ – △ – **?** – **?** – **?** –

Starting point is to identify the individual symbols by a simple code which is easier to work with than shapes – say 1 for a square, 2 for a circle and 3 for a triangle. The pattern is then rewritten in this (number) code:

1 – 2 – 3 – 1 – 2 – 3 – 1 – 2 – 3 – ? – ? – ?

It takes only a little study to see that the pattern is divided into groups of three which follow each other in the same manner, i.e. simply repeat themselves:

first group	second group	third group	fourth group
1 – 2 – 3	1 – 2 – 3	1 – 2 – 3	? – ? – ?

Typically the fourth group must be the same as the others,

i.e.   fourth group
                                    1 – 2 –3

or re-write original symbol form

These are, therefore, the missing symbols in the original question.

A little too simple and obvious? Then try to find the missing symbols in the following:

□ △ ▲ ▲ ○ □ ▲ ▲ △ ○ ■ ▲ △ △ – – – – –

First make a plan of the individual symbols involved.
These are

Square - □ **- call this 1.**

Four versions of a triangle – plain △ **– call this 2**

                              - solid ▲ **– call this $\bar{2}$ (it is still a triangle, but different)**

              small triangle △ **– call this (ii) ('little 2')**

              small solid ▲ **– which is logically $(\overline{ii})$**

              Circle - ○ **–call this 3**

The original pattern is then re-written as follows:

   1 (ii) $\bar{2}$ $\bar{2}$ 3   1 $(\overline{ii})$ 2 2 3 $\bar{1}(\overline{ii})$ 2 2 - - - - - -

This logically splits into groups of five symbols repeating themselves with *variations*. To get a better picture, arrange these groups in separate lines:

1st group	1	(ii)	$\bar{2}$	$\bar{2}$	3
2nd group	1	$(\overline{ii})$	$\bar{2}$	2	3
3rd group	$\bar{1}$	$(\overline{ii})$	2	2	–
4th group	–	–	–	–	–

We now have a picture of what is going on. All the *basic* symbols appear in the same order in each group, i.e. a square first, followed by a small triangle, then two large triangles, and lastly a circle. The only *variation* that occurs is that the solid symbols appear consecutively, moving one position to the left in each succeeding group. Hence we would be (logically) justified in completing the third and fourth groups as:

3rd group	$\bar{1}$	(ii)	2	2	3
4th group	$\bar{1}$	(ii)	2	2	3

Hence the missing symbols in the original are:

last in third group | fourth group

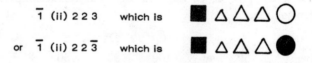

This is a logical solution not necessarily *completely* correct, but the best we can do with the available information. Certainly the first symbol in the third group is $\bar{1}$ or ■ ; but the last symbol in the third group *could* be $\bar{3}$ or ● (there is no evidence in the original information to say whether it is likely to be or not). If it is $\bar{3}$, then by the 'shift to the left' rule established, the fourth group could be either:

$\bar{1}$ (ii) 2 2 3   which is   ■ △ △ △ ○

or $\bar{1}$ (ii) 2 2 $\bar{3}$ which is   ■ △ △ △ ●

That, in fact, is the weakness of the *technique* of aptitude tests of this type. They are not necessarily capable of yielding only one answer by logical reasoning. There may be equally possible alternatives. Unfortunately, when set as test questions there is only one correct answer – the one the originator of the question has given.

This ambiguity is even more possible in solutions to questions or problems set which have to be answered by *abstract* reasoning. Which, for example, is the odd one out in this series of diagrams?

o|o   ⬨   %   ⊙⊙   %   ⬨
(a)   (b)   (c)   (d)   (e)   (f)

Reasoning would select (d) – the only one with the two circles on one side of the line. But (a) is also an odd one out on the reasoning that it is the only one with a vertical straight line. Also (c) is an odd one out on the reasoning that it is the only one which is a mathematical symbol (percentage).

Surprisingly – or perhaps not so surprisingly – abstract reasoning can be extremely puzzling, or even frustrating, to extremely intelligent people who tend to think 'logically'. The number of extremely clever children who give anything but the right answer to the following simple question is remarkable:

'If you were eating an apple and found half a maggot in it, what would you think?'

Numerical problems are much simpler for logical minds. One just looks for the pattern of 'formula' involved, in the same way as applying logic to patterns of symbols. Again the classic problem is to find the missing or next number(s) in a series of given numbers which are known to have been generated in some logical way. What, for example are the next numbers which follow logically from:

    50   40   100   90   150   ?   ?   ?   ?

To find the logical answer, write down the numbers as heads of columns lettered A, B, C, etc and see what relationship there is between columns:

	A	B	C	D	E	F	G	H	J
Number	50	40	100	90	150	?	?	?	?
Possible relationships	–	A–10 or D–A	$2 \times A$ or E/3 or A+50	A+B or C–10 or B+50	$3 \times A$ or A+C or C+50				

Notice how three repeated relationships have appeared:

(i) a '−10' relationship in alternate columns B and D (which could be anticipated as also following in columns F and H).

(ii) A 'multiplication' relationship in alternate columns C and E (which could be anticipated as also following in columns G and J)

(iii) A '+50' relationship in columns C, D and E (which could also be anticipated as $F=D+50$, $G=E+50$, $H=F+50$ and $J=G+50$).

Using the first relationship would give:

A	B	C	D	E	F	G	H	J
-	(A−10)		(C−10)		(E−10)		(G−10)	
50	=40	100	=90	150	*=140*	?	?	?

This gives the missing figures for F, but still leaving values for G, H and J unknown.

Using the second relationship gives:

A	B	C	D	E	F	G	H	J
-	-	2×A	-	3×A	-	4×A	-	5×A
50	40	100	90	150	140	*200*	-	*250*

This gives values for G and J. Now, knowing G, we can establish from the first relationship that the missing value for H should be $G-10=190$. Thus the missing figures are:

F	G	H	J
140	*200*	*190*	*250*

Using the third relationship gives:

A	B	C	D	E	F	G	H	J
-		A+50	B+50	C+50	D+50	E+50	F+50	G+50
50	40	100	90	150	*140*	*200*	*190*	*250*

This confirms the solution derived from (i) and (ii). In fact, it provides *all* the missing numbers, and so is a *complete* 'formula' for solution in itself. To check, see how the given values for B and A can also be shown to be consistent with this formula.

B = number before A + 50 = 40

Therefore number before A must be *−10*

A = number before the number before A + 50 = 50

Therefore the number before the number before A must be 0. In other words, if the series were extended leftwards from A, it would read:

			A	B	C .....
	0	−10	50	40	100 .....
check			(0+50	(−10+50	(50+50
			=50)	=40)	=100)

Incidentally the preceding number value − 10 could also be deduced from relationship (i); but neither relationship (i) nor (ii) would establish the number before that (0).

The *validity* of all the relationships has been proved by being repeated in the original information (the original series of numbers). The relationship in the original column D (D = A + B) appears only once, so there is no justification for it being applicable *again* in the series. If it were, by the same logic:

since   D = A + B
then   E should equal B + C or 40 + 100 = 140, which it is *not* (E = 150).

Thus this relationship is not repeatable, so is not valid.

As a further example of numerical reasoning, find the missing numbers in the following series:

2   4   4   16   16   ?   ?

write down in column:

	A	B	C	D	E	F	G
	2	4	4	16	16		
relationship	A	A^2	A^2	B^2	B^2	?	?

The *relationship pattern* to emerge is that following pairs of numbers are equal (i.e. B = C and D = E, so F = G); and that such pairs are the *square* of the preceding pair number. Hence F and G are 16^2 or 256.

CHAPTER 9

Introduction to Block Logic and Truth Tables

Many problems in logic can be readily solved by *block logic* using appropriate combinations of *logic functions*. The basic technique is very simple for only three different functions are involved:

NO (normally called *NOT*), *AND* and *OR*

The 'opposite working' or inverted forms of these are:

NOT NOT or *YES*; NOT AND called *NAND*; NOT OR called *NOR*

The attraction of this method is that you can represent each logic function by a rectangle with its function written in, and then think of this block as a switch which accepts a premise or *signal* applied to one side as an *input* and then either passes it through the block as an answer or *output* signal, or stops it. At the same time we can construct a *truth table* which shows all possible combinations of signals in and out.

A NOT logic block can accept only one input and has one output. It works like a switch which is normally closed. Thus a signal applied to the input *opens* the switch, so there is *no* output – Fig. 9.1. Information

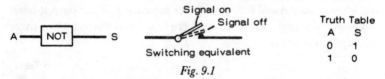

Fig. 9.1

fed to the block is rejected, i.e. NOT acceptable. This is also shown by the truth table as 'yes' or 'no' combinations. Conventionally 'yes' is written as 1 and 'no' as O to avoid possible confusion with other logic blocks, e.g. a YES logic *block* has a different truth table.

The inverted form of NOT is YES, corresponding to a switch which is *normally open*. On receipt of a signal the switch closes to pass the signal through as acceptable (YES) – Fig. 9.2.

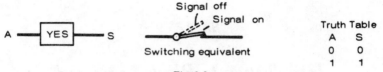

Fig. 9.2

Exactly the same result is obtained connecting two NOT devices in series, when the output of the first becomes the input to the second – Fig. 9.3. As the truth table shows, the final output has the same relationship to the input A as YES (NOT NOT logically is the same as YES). However, this uses two logic blocks instead of one – or two stages of analysis, if you like, instead of one. It is obviously simpler to provide a YES function with a single block (single device). Nevertheless the inverted forms of the logic blocks are extremely useful.

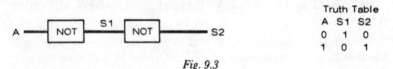

Fig. 9.3

AND and OR blocks (and their inverted forms) again have a single output but in this case can have two or more inputs – as many inputs as are necessary to accommodate the relative inputs in fact. For simplicity, we will consider just two inputs being used.

Fig. 9.4 then shows an AND logic block with inputs A and B, and output at S, together with its truth table. There is an output (S = 1) only when inputs A *and* B both = 1 (we have used 'S' for output here, designating a signal). In this case there is a signal output when there is an input at A *and* B.

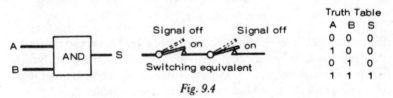

Fig. 9.4

The inverted form or NAND logic block is shown in Fig. 9.5. Here the truth table shows that when there is a signal input at A *and* at B there is no output (S = O). Thus anything that is to be inhibited when

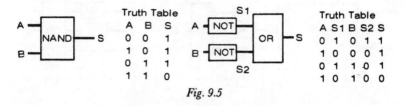

Fig. 9.5

signals are present at both A *and* B is satisfied by there being no output. All the other combinations of A and B signals produce an output. Again note that the same (NAND) logic function is also performed by an AND block with a NOT in each input line. (In this case calling for three blocks instead of one.)

Finally the OR block and its inverted form the NOR block are shown in Fig. 9.6. An OR block gives an output in the presence of a signal at A *or* B. A NOR block gives an output only when there is *no* signal A *or* B (neither A *nor* B).

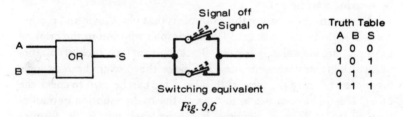

Fig. 9.6

Let's try a simple example of using logic blocks to solve a problem in logic, taking one of the examples used in Chapter 2. (There is a reason for this which will be explained later.) The example summary is repeated here to save looking it up.

(i) Brown, Green and White are each wearing a coloured hat
(ii) The colours of the hats are brown, green and white
(iii) Brown pointed out this fact
(iv) The one wearing the green hat then pointed out that none of the colours of the hats they were wearing was the same as their names.

Designate the hat colours B for brown, g for green and w for white. Take Brown, as he spoke first and this must be of some significance in the problem. Brown is wearing a brown hat *or* a green hat *or* a white hat – so connect b, g and w as inputs to an OR logic block – **Fig.**

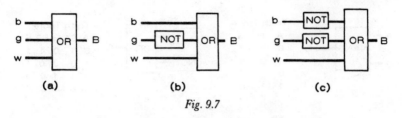

Fig. 9.7

9.7(a). Now from (iv), Brown cannot be wearing a green hat, so exclude this as a possible input by inserting a NOT in the g input line – Fig. 9.7(b).

From (iv) again Brown cannot be wearing a brown hat, so exclude this as a possibility by inserting a NOT in the b input line – Fig. 9.7(c).

The answer is now obvious from the final diagram alone – we don't even have to bother with constructing a truth table. Brown can only be wearing a white hat.

The reason for repeating this example is that block logic and mathematical logic (which was used for the simpler solution in the original example) are very closely related. Block logic, in fact, presents mathematical logic in diagrammatic form since they cover the *same* logic functions and have the *same* truth table. One can be used to check the other. If a question arises as to the validity of an equation in mathematical logic (Boolean algebra), it can be 'spelt out' in the form of block logic to see if it makes the proper sense.

There is one important difference, though. The use of block logic, especially for more complicated problems, can lead to redundant blocks being introduced, i.e. more logic blocks than are strictly necessary. Also redundancies are not always easy to spot or eliminate. This is not necessarily important using block logic for solutions in deductive logic, but it is if block logic is being used to design control *circuits*. It means that the final circuit ends up by using more switching devices (logic blocks) than are strictly necessary. If the same problem is solved in mathematical logic (Boolean algebra) it is readily possible to simplify equations and thus eliminate all redundancies.

Both block logic and Boolean algebra deal with logic in terms of basic logic functions, and in an *exact* way (as opposed to drawing a conclusion from premises in deductive logic) – the one in a diagrammatic way, and the other with mathematical equations. Equally, because they complement each other, when circuit designs for example are worked out in mathematical logic, the final equation can then be redrawn in block logic form to show the number and types of logic devices required.

Equally, if designed in block logic, the final diagram could then be rendered in the equivalent mathematical logic equation which is then studied to see if it can be simplified.

The main use of both block logic and mathematical logic, in fact, is in functional circuit design - electrical circuits (ranging from quite elementary switching circuits to those for microprocessors and computers), and control circuits in hydraulics and pneumatics. Surprisingly, neither is used as much as it could be in solving 'non-functional' logic problems. Think about it before you automatically use deductive logic for dealing in premises. If premises can be turned into 'signals' operating on a yes-no basis (true or false), positive solutions can be obtained in terms of NOT, OR and AND logic.

Equally, if you prefer to work with diagrams, then construct a truth table for the block logic involved. In many cases you can use the simpler, more easily understood simple logic diagrams described in Chapter 5. The other alternative is Venn diagrams (Chapter 4), although here there is a greater possibility of making mistakes until you become thoroughly familiar with this type of diagram.

Block Logic and Circuit Design

Block logic is a useful and readily mastered tool for circuit design involving switching or control elements. For electrical circuits, each logic function AND, OR, NOT can be performed by a switch (or combination of switches). In hydraulic or pneumatic control circuits, similar functions can be performed by control valves. Circuits can thus be designed with logic blocks, and then finalized by substituting the appropriate switches or valves for each logic block.

Let's start with a very simple example. Suppose it is required to design a circuit to switch something on and off from two separate positions A and B. Drawn as a solution in block logic this simply involves an OR function - Fig. 9.8. This is easily translated into two separate switches, one at position A and one at position B, as in the second diagram.

Suppose, now, it is necessary to be able to switch 'on' and 'off' from *either* station. The first circuit will not work. If switched on at A, for

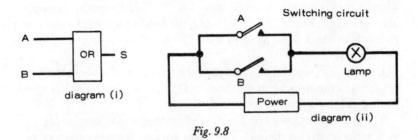

Fig. 9.8

instance, and left on, it cannot be switched off at B, and vice versa. Additional logic is required in this case, as shown in Fig. 9.9, with the corresponding switching circuit shown in the second diagram. Now if switch A is left 'on', for example, switch B will be in its 'off' position and the circuit completed through its 'off' position (i.e. B̄ state). Moving

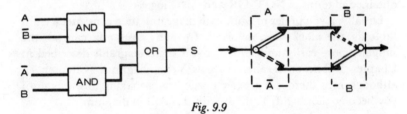

Fig. 9.9

switch B to its 'on' position will then break the circuit (with switch A still 'on'), and thus switch off the *circuit*. The same occurs if switch B is left 'on', with switch A 'off'. Switching A 'on' will switch the *circuit* 'off'. In fact the two switches do not have any specific 'on' or 'off' positions. They are merely acting as 'gates' to perform a combination of AND and OR functions.

It is also possible to produce the same working with a different arrangement of logic blocks, as shown in Fig. 9.10. This diagram

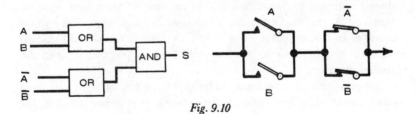

Fig. 9.10

expresses the required logic in the form – either A OR B 'on' AND (at the same time) B OR A 'off'. In fact, we might have thought of planning the block logic diagram first in this way.

An interesting fact now emerges. The switching equivalent of this block logic solution, shown in the right-hand diagram, is considerably more complicated than that for the alternative solution. In other words the solution in OR logic is more complicated as regards switches than the solution in AND logic. The reason for this is that AND logic operates in *series* fashion and OR logic works in *parallel* fashion. In terms of mechanical switching, series working is less complicated than parallel working. Circuit designs in AND logic, therefore, are normally simpler than designs in OR logic when translated into mechanical switches. It does not follow, however, that this is also true when the switching is done by *electronic* 'gates'.

As a further matter of interest, solutions to the above problem could equally well have been derived from truth tables or Boolean algebra. The requirement, expressed in Boolean algebra is:

$$A\bar{B} + \bar{A}B = 1$$

(i.e. A 'on' with B 'off' OR A 'off' with B 'on' gives an output). This would translate in logic block form as in Fig. 9.9.

Looking at the requirement the other way round, i.e. the combinations that do *not* produce an output, corresponding to the complete circuit being switched 'off'), the Boolean equation would be:

$$AB + \bar{A}\bar{B} = 0$$

which for 'switch on' working has to be inverted and becomes:

$$\overline{(AB + \bar{A}\bar{B})} = 1,$$

which by applying de Morgan's theorem becomes:

$$(A + B) \cdot (\bar{A} + \bar{B}) = 1$$

This would translate in block logic form as in Fig. 9.10.

More Complex Circuits

The control problem solved in Fig. 9.11 by block logic is a little more complicated. The device being controlled is a machine which can be started from either of two operator positions A or B. Before the machine can be started, however, it is necessary that the component

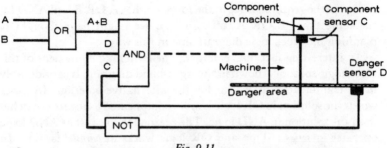

Fig. 9.11

it is handling is in position (as detected by the component sensor C), and there is nobody standing in the danger area. Sensor D generates a danger signal if anyone *is* in this area.

The logic required is A *OR* B *AND* C *AND* NOT-D. A and B are connected to an OR block, and the output of this block provided one input to the AND block. C is also connected to the AND block, giving a second input to this block. D is connected to the AND block via a NOT block and is the third input to this block.

There will be an output from the AND block, switching the machine on, *only* when there is an input signal at *each* of the three input lines to the AND block. If either operator A or B switches the machine on, the machine will only start if there is also an input signal on the C line (component in position) and *no* signal on the D line. The NOT in this line will then invert this '0' signal into a '1', providing the third signal input to the AND block. If there *is* a signal on the D line (someone standing in the danger area), the NOT in this line will invert that '1' into a '0' – i.e. there will be no input from this line into the AND block. Consequently there will be no output from the AND block and the machine will not start. In Boolean algebra this is presented quite simply as:

$$(A+B).C.\overline{D}=1$$

In practice – and with experience – circuit designs of this type are commonly developed in a specific *type* of logic. The solution in Fig. 9.11, for instance, uses a mixture of OR and AND logic elements – one of each in this case, but in a more complex control circuit there may be several of each required.

Depending on the availability of actual control elements, or possibly looking for the simplest solution using the minimum number of logic elements, alternative solutions are often worth looking at. Virtually any problem of this type can be solved either in AND logic *or* OR logic; or the inverted forms NAND *or* NOR logic. In this case only one *type* of logic element will be required, together with NOT elements for essential inversion.

Fig. 9.12 shows the same problem as Fig. 9.11 solved in NOR logic. Specifically, this involves using two NOR devices and five NOT devices – seven devices in all, but we have got everything in NOR

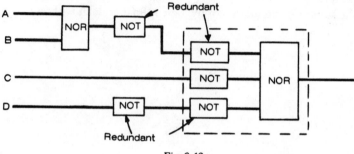

Fig. 9.12

logic. Also we can simplify this diagram. Two NOTs in the same line simply change a signal back to what it was originally – they are *redundant* elements which can be eliminated. On this basis we can eliminate four devices from the original and end up with just three devices performing the same function, still all in NOR logic – Fig. 9.13.

Try working out further alternative solutions in AND logic, OR logic and NAND logic only. (Remember you will have to use NOT devices as well.)

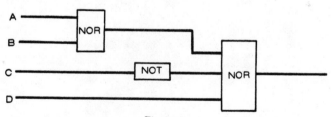

Fig. 9.13

The main limitation with circuit design by block logic is that the more complex the problem to be solved the greater the number of logic blocks (devices) required and the more likely redundancies will be introduced. Also it may not be easy to spot all the redundancies and elimate them. Using Boolean algebra it is much easier to simplify the equations. Even this has its limitations when it comes to the design of complex pneumatic and hydraulic circuits where Karnaugh maps can prove much more effective, but represent a new logic technology to master.

The subject of circuit design by logic is further dealt with in Chapters 11, 12 and 13.

CHAPTER 10

Algorithms

An algorithm is a modern way of plotting logic problems (the term was not invented until the 1960s) which basically presents a chart accommodating all the information relative to a problem and leads one exactly through the paths necessary to arrive at the logical result. Basically, in fact, it is simply an extension of block logic, given a more sophisticated name. It was originally devised as a means of plotting the strategy of solving problems using computers. It is now widely used for designing flow-charts, making it possible for people to arrive at solutions on a yes/no basis (like the working of a computer), with the advantage that it can be completely non-mathematical. The scope in this respect is considerable. Algorithms can be applied equally well in general problem solving, commerce, industry, ergonomics, medicine, finance, even political and military strategy. There are, equally, mathematical algorithms. All have one thing in common. They are directed to finding solutions – not teaching or learning. Algorithms do not help understand a problem like a computer programme; therefore, the solutions they give are only as good as the person designing the complete algorithm.

Non-mathematical algorithms on individual facts or statements relevant to the subject rendered as questions, which become in effect 'gates' in the flow path. Each 'gate' provides two exits – one for a 'yes' answer and one for a 'no' answer, this procedure being followed through the algorithm until the final answer part(s), which is then the logical solution to the answers made at the various question parts. Rather than a single answer, too, an algorithm commonly contains a number of different answers since it is designed to cover all contingencies, i.e. a 'yes' or a 'no' at all of the gates. A simple example will make this clear:

A club has the following rules regarding subscriptions:

(i) Entrance fee is £50 for adult members; £20 for persons under 20 years on January 1st.

(ii) Annual subscription rates payable on September 1st each year are:

Full voting member	£50
Senior (non voting) member	£35
Junior (non voting) member	£20

An algorithm drawn up on this basis would look like this:

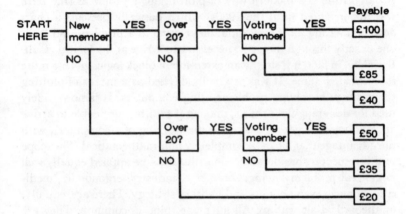

This chart plots all the possibilities with logical paths through to the amount payable. It can provide *immediate* solutions relative to any applicant, new or existing member, whether over or under 20, and whether voting or not. In the same way it could be extended to cover other club subscription rules. Try re-drawing it to take into account as well:

(iii) Members joining after March 31st in any year pay only half that year's subscription.

(iv) Family membership (two persons from the same family) can have 25 per cent reduction on subscription rate.

(v) Existing members over 65 years old, with at least five years paid-up membership, can have a reduced membership of £40 for full membership, or £20 for individual (non-voting) membership.

Algorithms of this type can readily be prepared to cover almost all types of subjects involving variable factors which can be dealt with on a yes/no basis. Once prepared, they then cover all possibilities involving these factors, with a logical route to the correct answer in each case. This can save a lot of time and effort compared with working out the answer each time for individual cases.

The main problem to arise in designing algorithms – apart from ensuring that *all* the factors are included and presented on a yes/no basis – is one of arrangement. Ideally – but not necessarily – all the 'yes' paths should come out the same way, and all the 'no' paths in the same (but different) way. Thus in the example drawn, a 'yes' automatically reads straight across to the right. A 'no' reads downwards and then across on a different path.

The other difficulty, particularly when a large number of different 'boxes' have to be used is to avoid *crossing* paths, as this can be misleading. Also the final 'answering' boxes should be made distinct from the others (e.g. with a bolder outline).

Drawing up the basis of an algorithm is quite a simple exercise in block logic once you have got the facts broken down into simple questions. Planning and drawing up the final algorithm, however, can be something of an art to get it into its best presentable form. It may even be necessary if the algorithm tends to become excessively complicated to stop at certain answer part(s) which, instead of being answers, are new starting points for a separate algorithm.

Algorithms and Truth Tables

An algorithm can also be presented in the form of a *truth table*, from which relevant solutions can be drawn. To demonstrate this we will use the same example as before, dealing with club subscriptions, writing out each question as a line and covering all possible combinations of answers in the following columns of the table:

	1	2	3	4	5	6	7	8
New member?	YES	YES	YES	NO	NO	NO	YES	NO
Over 20?	YES	YES	NO	NO	NO	YES	NO	YES
Voting member	YES	NO	NO	NO	YES	YES	YES	NO
Voting £100	√							
£85		√						
£50						√		
£40			√					
£35								√
£20				√				

The truth table is more explicit in that it analyses *all* the possible combinations of the variable factors (answers to the questions). There are three questions, each with two possible answers ('yes' or 'no'), which in fact gives us *eight* possible combinations, as given by the eight columns in the truth table.

In this particular case the combination in columns 5 and 7 cannot apply by the club rules since the person is under 20 and thus a junior member, and a junior member is a non-voting member. With a different set of questions *all* possible combinations may be relevant, in which case there would be eight separate solutions, not six as in this example. Plotting a truth table, in fact, is a good cross-check on the completeness of the algorithm.

CHAPTER 11

Introduction to Boolean Algebra

The simplest approach to solving logic problems in a mathematical way is the use of Boolean algebra, which is certainly not as frightening as it sounds. It is like ordinary algebra in many ways, but certainly much simpler because it is concerned with only two possible states of each individual subject. These can be evaluated as true or false, yes or no, in the case of general subjects (e.g. premises); or on or off, go or stop, in the case of signals. (The usual interpretation with signals is 'I' for 'on' and 'O' for off').

At first it may look more complicated than ordinary algebra because the mathematical signs are used in a different sense, $\cdot$ for AND; $+$ for OR. Thus $A \cdot B$ means A *and* B (not A *times* B); and $A + B$ means A *or* B (not A plus B).

The only other symbol used is a $^-$ over a letter, meaning an inversion or logic NOT. Thus $\bar{A}$ means *not* A, $\bar{B}$ means *not* B, and so on.

It then becomes obvious that Boolean algebra works in terms of NOT, AND and OR logic, or the inverted forms YES, NAND and NOR.

By far the simplest way of understanding Boolean algebra is to think in terms of switching circuits applied to the basic equations for logic functions, as in Chapter 9.

NOT then becomes a *normally closed* on-off switch within the possibility of either a negated or positive output. The two Boolean algebraic forms of NOT are thus:

	A	S
$A = \bar{S}$	1	0
consistent with the truth table		
$\bar{A} = S$	0	1

YES then becomes a *normally open* on-off switch. The two Boolean algebraic forms of YES are thus:

	A	S
$A = S$	1	1
consistent with the truth table		
$\bar{A} = \bar{S}$	0	0

AND is the equivalent of two *normally open* switches in series. Here the presence of an input signal at either A or B closes its respective switch to pass on the signal to the output, but there is no output until *both* switches are closed (signals A *and* B are present):

A·B = S   consistent with the truth table

A	B	S
0	0	0
1	0	0
0	1	0
1	1	1

NAND is not quite so easy to describe in simple switching terms. Logically it should be two switches in series operating in the inverted mode to AND. In fact it works in the manner of one *normally closed* switch which needs signal inputs at both A and B to open it. With no input at either A or B the switch remains closed, providing a through path to the output and an output signal (consider this signal as coming from a separate source). With an input at A the switch still remains closed, so there is still an output. Similarly with an input at B only. With inputs at both A and B the switch is opened, breaking the circuit and changing the output to 0.

$\overline{A \cdot B}$ = S   consistent with the truth table

A	B	S
0	0	1
1	0	1
0	1	1
1	1	0

OR is straightforward again – two normally open switches in *parallel*. There is an output when either switch A *or* switch B is closed (by signal input at A or B, respectively. Also there is obviously an output when both switches are closed, as the truth tables shows:

A + B = S   consistent with the truth table

A	B	S
0	0	0
1	0	1
0	1	1
1	1	1

There is, however, another form of OR known as an EXCLUSIVE OR. This *excludes* the possibility of an output being generated when signals are present at *both* inputs. That is, the production of an output is exclusive to either A or B input signals being present.

NOR again is consistent with parallel switches giving an output (both switches closed) when there is not a signal input at A or B. This time either signal A or signal B (or both together) will open *both* switches, giving no output.

$$\overline{A+B}=S$$ consistent with the truth table

A	B	S
0	0	1
1	0	0
0	1	0
1	1	0

Note here that, with the exception of NOT and YES which can only have one (input) designation, all of the logic functions can accommodate any number of input signals A,B,C,D, etc, as necessary which can be expressed in terms of Boolean algebra. These merely extend the size of the truth table *in the same truth sense*. For example, here is the AND truth table extended to accommodate four inputs A, B, C, D.

A	B	C	D	S
0	0	0	0	0
0	0	0	1	0
0	0	1	0	0
0	0	1	1	0
0	1	0	0	0
0	1	0	1	0
0	1	1	0	0
0	1	1	1	0
1	0	0	0	0
1	0	0	1	0
1	0	1	0	0
1	0	1	1	0
1	1	0	0	0

A	B	C	D	S
1	1	0	1	0
1	1	1	0	0
1	1	1	1	1

This illustrates the basic rule that for N inputs there are 2^N possible states or combinations, represented by N lines in a truth table. Thus in this case there are sixteen lines (different states) possible with the switching equivalent of sixteen on – off switches connected in series. There is an output $(S=1)$ only when all sixteen switches are closed (i.e. $A=1$, $B=1$, $C=1$ and $D=1$). The truth table represented this more compactly than drawing out all the individual switches connected together.

Inverting the Function

One thing may – or may not – have become apparent from the logic equations explained above: inverting a function changes the *nature* of the function. Thus AND which is obviously an AND function $(A \cdot B$ or A *and* B), becomes an OR function when inverted to NAND (not A *or* not B). This is a very important rule in manipulating Boolean equations, where it is often convenient, or necessary, to invert functions to obtain an optimum solution (eliminate redundancies, for example). Specifically, too, it enables all expressions in an equation to be in the same *type* of logic.

Suppose we want to work entirely in OR logic. This means that all the terms used in an equation need to be expressed in the form $A + B$, i.e. the $+$ (OR) sign is used throughout the equation.

Starting with the NOR function, $\bar{A} \cdot \bar{B} = S$, this is in AND logic (as shown by the · sign). By the rule mentioned above, inverting it will change it into OR logic:

thus $\bar{A} \cdot \bar{B} = S$ becomes $\bar{\bar{A}} + \bar{\bar{B}} = \bar{S}$ by inverting each term and changing the sign

This now contains double inversions on A and B, which is bad logic. If inverted once (in the original form) and then inverted again (in the conversion to OR logic), any subject ends up the right way up – or in this case $\bar{\bar{A}} = A$ and $\bar{\bar{B}} = B$, so NOR converted to OR logic now becomes:

$$\bar{\bar{A}} + \bar{\bar{B}} = A + B = \bar{S}$$

We still have the output inverted ($\bar{S}$), which may or may not be convenient. If not, we can invert the whole expression again:

when $A + B = \bar{S}$ becomes $\overline{A + B} = S$

Thus we started with not-A *and* not-B = S (or NOR in AND logic) and end up with not-A *or* B = S (or NOR in OR logic, which was what we were after).

At this point there is something very important to note. When we first inverted the equation we changed the *sign* from · (AND) to + (OR), according to the rule of Boolean algebra which demands this, i.e. theorem (xii). When we did the last inversion, we did *not* change the sign. This is because we inverted $A + B$ as a whole, which can be considered as putting it in brackets $(A + B)$ and treating it as a single term. Again this is consistent with theorem (xii). (The theorems governing Boolean algebra are summarized together on pages 94–5 for convenience of reference.)

Ultra-Simplified Symbolic Logic

Now let us look at an ultra-simple symbolic method of handling deductive logic, in which we use just the inversion sign ¯ borrowed from Boolean algebra, together with letter symbols. It is not Boolean algebra, and indeed the validity of the method is questionable – but it works. Take the following as an example:

Premises: all dogs are animals
no animals can fly
Conclusion, by deductive logic: no dogs can fly.

Symbolic technique: designate all dogs D, animals A, and fly F

first premise becomes $D = A$
second premise becomes $\bar{A} = F$
   Therefore if $\bar{A} = F$
                    $A = \bar{F}$
thus $D = A = \bar{F}$

(All dogs cannot fly, which is the same as no dogs can fly.) Try this technique out for yourself using the examples in the chapter on Deductive Logic (Chapter 2). In fact, we have already used this quick method to provide a second solution to the 'coloured hats' problem in Chapter 2. Let's see how this same solution can be worked out correctly in Boolean algebra, using the same notation as before – B for Brown and b, g and w for the brown, green and white hats respectively.

The start is exactly the same as before:

$$B = b + g + w$$

also because the man who spoke after Brown was wearing a green hat, Brown cannot be wearing a green hat. Nor can Brown be wearing a hat the same colour as his name, so

$$B = \bar{g} \cdot \bar{b}$$

Last time we jumped straight to the answer by assuming that $\bar{g}$ and $\bar{b}$ cancelled out g and b in the first equation. This is not justifiable in Boolean algebra since there are not *two* separate equations involved, only *one* combining the two propositions, viz:

$B = b + g + w$ *and* $\bar{g} \cdot \bar{b} \cdot$ or
$B = (b + g + w) \cdot \bar{g} \cdot b$ which expands to
$B = b \cdot \bar{b} \cdot \bar{g} + g \cdot \bar{g} \cdot b + w \cdot \bar{g} \cdot \bar{b}$ and then groups in OR logic as
$B = (b \cdot \bar{b} \cdot \bar{g}) + (g \cdot \bar{g} \cdot \bar{b}) + (w \cdot \bar{g} \cdot \bar{b})$
From theorem (i) $b \cdot \bar{b} = 0$
$\qquad\qquad\qquad g \cdot \bar{g} = 0$
thus $B = (0 \cdot \bar{g}) + (0 \cdot \bar{b}) + w \cdot \bar{g} \; \bar{b}$
From theorem (i) $0 \; \bar{g} = 0$
$\qquad\qquad\qquad 0 \; \bar{b} = 0$
Hence $B = w \cdot \bar{g} \cdot \bar{b}$
$\qquad\quad$ = white, NOT green, NOT brown.

Exactly the same answer as the short cut solution gave!

Boolean Algebra Theorems
$\quad$ (i) Since 0 has no value other than zero
$\qquad\qquad A \cdot 0 = 0$ (AND function)
$\quad$ (ii) Since 1 has no other value than 1 (positive)
$\qquad\qquad A \cdot 1 = A$ (AND function)
$\quad$ (iii) Since 0 has no other value than zero
$\qquad\qquad A + 0 = A$ (OR function)

(iv) Since 1 has no other value than 1 (positive)

$$A + 1 = 1 \quad \text{(OR function)}$$

(v) $\quad A \cdot A = A \quad \text{(AND function)}$

(vi) $\quad A + A = A \quad \text{(OR function)}$

(vii) $\quad A \cdot \bar{A} = 0$

(viii) $\quad A + \bar{A} = 1$

(ix) $\quad \bar{\bar{A}} = A$ (double inversion returns the function to its original state)

(x) The order of individual functions is immaterial

$$A \cdot B = B \cdot A$$
$$A + B = B + A$$

(xi) The order of grouping of functions is immaterial

$$A \cdot (B \cdot C \cdot) = (A \cdot B \cdot) \cdot C$$
$$A + (B + C) = (A + B) + C$$

(xii) Inversion of all signals changes the sign (de Morgan's theorem)

$$\overline{A \cdot B \cdot C} = \bar{A} + \bar{B} + \bar{C}$$
$$\overline{A + B + C} = \bar{A} \cdot \bar{B} \cdot \bar{C}$$

(xiii) For simplification of equations . (AND) is treated as a multiplication sign and + (OR) as an addition sign, as in ordinary algebra

$$A \cdot (B + C) = A \cdot B + A \cdot C$$

(xiv) Expansion of equations treated as in ordinary algebra

$$(A + B) \cdot (C + D) = A \cdot C + A \cdot D + B \cdot C + B \cdot D$$

(xv) Redundancies can be eliminated, e.g.

$$A + (A \cdot B) = A$$
$$A + (\bar{A} \cdot B) = A + B$$

This last theorem may not be obvious, but can be proved by its truth table:

A	B	S
0	0	0
1	0	1
1	1	1
0	1	1

This shows that $A = 1$ makes $S = 1$ OR $B = 1$ makes $S = 1$ whether $A = 1$ or $A = 0$. Thus $\bar{A}$ is redundant in the equation.

Minterms and Maxterms

Boolean algebra and truth tables are closely related, as we have seen. One can always be expressed in terms of the other. However the logic involved may be a mixture of AND, OR and inversions of AND and OR. Assuming that we start with a truth table, for a systematic approach to combinational logic the truth table needs to be expressed in a Boolean expression with a particular form, i.e. AND logic or OR logic. In AND logic the Boolean expressions are called *minterms*; and in OR logic the Boolean expressions are called *maxterms*. The two are complementary, as solutions can be worked in either. Here, for example, is the truth table for covering three binary variables A, B and C and the corresponding Boolean minterms and maxterms:

A	B	C	minterm	maxterm
0	0	0	$\bar{A}\cdot\bar{B}\cdot\bar{C}$	$\bar{A}+\bar{B}+\bar{C}$
1	0	0	$A\cdot\bar{B}\cdot\bar{C}$	$A+\bar{B}+\bar{C}$
0	1	0	$\bar{A}\cdot B\cdot\bar{C}$	$\bar{A}+B+\bar{C}$
1	1	0	$A\cdot B\cdot\bar{C}$	$A+B+\bar{C}$
0	0	1	$\bar{A}\cdot\bar{B}\cdot C$	$\bar{A}+\bar{B}+C$
1	0	1	$A\cdot\bar{B}\cdot C$	$A+\bar{B}+C$
0	1	1	$\bar{A}\cdot B\cdot C$	$\bar{A}+B+C$
1	1	1	$A\cdot B\cdot C$	$A+B+C$

This is a particularly useful technique for the design of combinational circuits (e.g. electrical circuits or hydraulic or pneumatic control circuits). If the relationship (i.e. significance of the 'switching' points A, B and C) are stated in minterms, the circuit can be designed using digital devices having an AND function, plus NOT for inversions. If the relationship is in maxterms, then the circuit can be designed using OR devices, plus NOT for inversions.

Repeating this table but this time using only minterms and adding the output:

A	B	C	minterm	output (S)
0	0	0	$\bar{A}\cdot\bar{B}\cdot\bar{C}$	0
1	0	0	$A\cdot\bar{B}\cdot\bar{C}$	1
0	1	0	$\bar{A}\cdot B\cdot\bar{C}$	0
1	1	0	$A\cdot B\cdot\bar{C}$	1
0	0	1	$\bar{A}\cdot\bar{B}\cdot C$	1
1	0	1	$A\cdot\bar{B}\cdot C$	0
0	1	1	$\bar{A}\cdot B\cdot C$	0
1	1	1	$A\cdot B\cdot C$	1

Here it is seen that there is an output if any of $A\cdot\bar{B}\cdot\bar{C}, A\cdot B\cdot\bar{C}, \bar{A}\cdot\bar{B}\cdot C$ and $A.B.C$ is 1. The corresponding Boolean equation is then:

$$S = A\cdot\bar{B}\cdot\bar{C} + A\cdot B\cdot\bar{C} + \bar{A}\cdot\bar{B}\cdot C + A\cdot B\cdot C$$

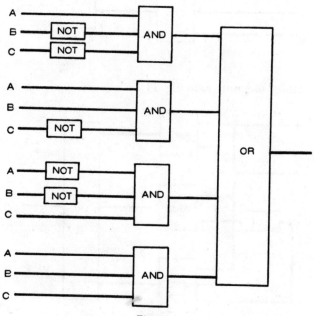

Fig. 11.1

Fig. 11.1 shows this as a complete circuit, using logic elements. It is an unnecessarily complicated circuit for it contains a number of redundancies. The original equation is capable of simplification (using the relevant Boolean algebra theorems) to

$$S = A \cdot B + A \cdot \check{C} + \bar{A} \cdot B \cdot C$$

This gives the simple circuit of Fig. 11.2.

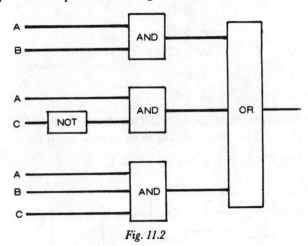

Fig. 11.2

Or as a 'wiring' diagram, as in Fig. 11.3

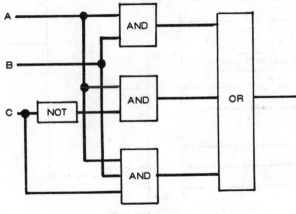

Fig. 11.3

CHAPTER 12

More Logic Equations

One particularly useful feature of Boolean algebra is that it is possible to express all the logic functions involved in a *single* logic. Problems involving OR and AND, for instance, can be expressed entirely in either OR logic or AND logic – or NOR or NAND logic. Working out how this can be done is also useful practise in manipulating Boolean equations and applying the rules laid down in the theorems.

Everything in AND Logic

In AND logic the basic expression is $A \cdot B = S$. Thus to be in AND logic every piece of logic must be expressed in the $\cdot$ form.

An OR function is changed into AND logic as follows. Start by writing down the basic OR function:

$$A + B = S$$
$$\text{invert} \quad \bar{A} \cdot \bar{B} = \bar{S}$$

Note this changes the sign from OR to AND, but the output is inverted. Invert again to obtain a positive output:

$$\overline{\bar{A} \cdot \bar{B}} = \bar{\bar{S}} = S$$

Thus $\overline{\bar{A} \cdot \bar{B}} = S$ is the OR function in AND logic.

The NOR function is an inverted OR, and so is already in AND form with a positive output. No conversion is necessary:

$$\bar{A} \cdot \bar{B} = S$$

The basic NAND function is in OR logic, viz:

$$\bar{A} + \bar{B} = S$$
$$\text{invert to give} \quad \underline{A \cdot B = \bar{S}}$$
$$\text{invert again for a positive output} \quad \overline{A \cdot B} = \bar{\bar{S}} = S$$

Thus $\overline{A \cdot B} = S$ is the NAND functions in AND logic.

Everything in OR Logic

In this case every piece of logic needs to be expressed in $+$ form.

Starting with the AND function:

$$A \cdot B = S$$

invert to put into OR logic:

$$\bar{A} + \bar{B} = \bar{S}$$

invert again for a positive output:

$$\overline{\bar{A} + \bar{B}} = \bar{\bar{S}} = S$$

Thus $\overline{\bar{A} + \bar{B}} = S$ is the AND function in OR logic.
The NAND function is already in OR logic, i.e.

$$\bar{A} + \bar{B} = S$$

The NOR function is in AND logic, i.e.

$$\bar{A} \cdot \bar{B} = S$$

invert to put into OR logic

$$\bar{\bar{A}} + \bar{\bar{B}} = \bar{S}$$

which is the same as $A + B = \bar{S}$
invert again to get a positive output

$$\overline{A + B} = S$$

Thus is the NOR function in OR logic.

Everything in NAND Logic
In this case since NAND is inverted AND, every piece of logic needs to be expressed in $\overline{\cdot}$ form.
Starting with the AND function

$$A \cdot B = S$$

invert to get the NAND form

$$\overline{A \cdot B} = \bar{S}$$

invert again to get a positive output

$$\overline{\overline{A \cdot B}} = \bar{\bar{S}} = S$$

This is the AND function in NAND logic.
The OR function is $A + B = S$

invert to give

$$\bar{A} \cdot \bar{B} = \bar{S}$$

invert again to get the $\overline{\quad \cdot \quad}$ form

$$\overline{\bar{A} \cdot \bar{B}} = \bar{\bar{S}} = S$$

Thus $\overline{\bar{A} \cdot \bar{B}} = S$ is the OR function in NAND logic.

The NOR function is $\quad \bar{A} \cdot \bar{B} = S$
invert as a whole into the $\overline{\quad \cdot \quad}$ form

$$\overline{A \cdot B} = \bar{S}$$

invert again to get a positive output

$$\overline{\overline{A \cdot B}} = \bar{\bar{S}} = S$$

Thus $\overline{\overline{A \cdot B}} = S$ is the NOR function in NAND logic.

Everything in NOR Logic

This time everything needs to be expressed in $\overline{\quad + \quad}$ form.

The AND function is $\quad A \cdot B = S$
invert to give

$$\bar{A} + \bar{B} = \bar{S}$$

invert again to get into $\overline{\quad + \quad}$ form

$$\overline{\bar{A} + \bar{B}} = \bar{\bar{S}} = S$$

Thus $\overline{\bar{A} + \bar{B}} = S$ is the OR function in NOR logic

The NAND function is $\quad \bar{A} + \bar{B} = S$
invert as a whole to get into $\overline{\quad + \quad}$ form

$$\overline{A + B} = \bar{S}$$

invert again to get a positive output

$$\overline{\overline{A + B}} = \bar{\bar{S}} = S$$

Thus $\overline{\overline{A + B}} = S$ is the NAND function in OR logic.

Circuit Design by Equations

Examples of circuit design by block logic have already been given in Chapter 9 where, too, Boolean algebraic equations were used to check the results. (You probably skipped that part if you had not already read Chapter 11.) All combinational control circuits can be expressed in terms of block logic or pure equations – or the requirements equally well written out in the form of a truth table (which, of course, will be the same for the block logic and algebraic solutions). Provided you are happy working with Boolean algebra, solution by equation is usually simpler and quicker, and redundancies can be eliminated in the process.

All combinational control circuits are similar in that they provide an 'on' signal when, and only when, a particular *combination* of other signals are present. Thus the example given here is basically the same sort of problem as solved by block logic in Chapter 9, but a little more complicated in involving a combination of five different signals. Circuit design will be worked out in Boolean algebra on a logic basis, but also illustrated by block logic diagrams as 'pictorial proof' of what the equations mean.

The control problem, shown in Fig. 12.1, is this. A space rocket site incorporates the following features. At the control centre, either technician A or B can give the final signal to start ignition. The rocket is fuelled up, ready to go, but it must be impossible for A or B to launch the rocket until all the monitoring signals are in a 'go' state. These monitoring signals are derived as follows.

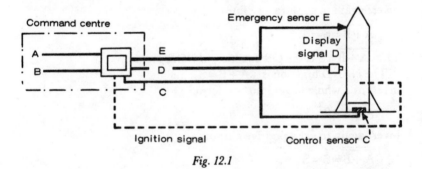

Fig. 12.1

A control sensor (call it C) gives a continuous signal indicating that the rocket is sitting properly, ready to go. A display screen (call it D) monitors the internal systems, generating a ready-to-go signal when all functions are operating correctly. A further emergency sensor (call it E) on the tower only gives a signal if something else is wrong and the launch must not take place.

In terms of logic we have the following requirement (or logic equation):

A OR B AND C NAND D AND NOT E = S (ignition signalled)

which in symbols is:

$$A + B \cdot C \cdot D \cdot \bar{E} = S$$

This is in mixed logic (OR and AND logic, together with NOT). A control circuit, based on logic elements, would be as in Fig. 12.2.

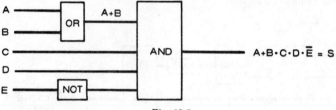

Fig. 12.2

It could equally well be rendered in terms of AND devices only (AND logic), together with NOT.

Starting with the original equation:

$$A + B \cdot C \cdot D \cdot \bar{E} = S$$

invert once

$$\bar{A} \cdot \bar{B} + \bar{C} + \bar{D} + \bar{\bar{E}} = \bar{S}$$

invert again to get into AND form and also give a positive output:

$$\overline{\bar{A} \cdot \bar{B}} \cdot \bar{\bar{C}} \cdot \bar{\bar{D}} \cdot \bar{\bar{\bar{E}}} = \bar{\bar{S}} = S$$

This could be rendered in logic element form as shown in Fig. 12.3,

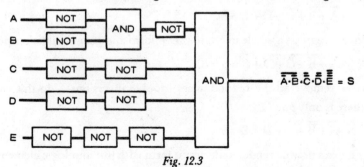

Fig. 12.3

but with obvious redundancies (NOT devices following each other for double inversions and effectively cancelling each other out).

This is also obvious from the equation. Double inversions cancel each other out, so the simplest form of the equation in AND logic becomes (by remaining double inversions)

$$\overline{\overline{A} \cdot \overline{B}} \cdot C \cdot D \cdot \overline{E} = S$$

The requirements for this circuit are shown in Fig. 12.4. It has saved six (NOT) devices.

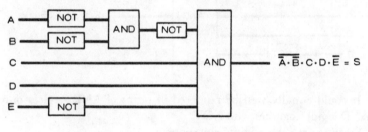

Fig. 12.4

Perhaps it could be made even simpler using another form of logic? Certainly it can. Suppose we try NOR logic. This means reading the terms of the equation in the form $\overline{+}$

Starting with the original equation:

$$A + B \cdot C \cdot D \cdot \overline{E} = S$$

invert, treating $A + B$ as a whole to retain the $+$ form:

$$\overline{A + B} + \overline{C} + \overline{D} + \overline{\overline{E}} = \overline{S}$$

invert again as a whole to retain the $+$ form and get a positive output:

$$\overline{\overline{A + B} + \overline{C} + \overline{D}} + \overline{\overline{E}} = \overline{\overline{S}} = S$$

Now remove all the redundancies (double inversions) – in this case there is only one ($\overline{\overline{E}}$):

$$\overline{\overline{A + B} + \overline{C}} + \overline{D} + E = S$$

This can now be rendered in circuit form with just four logic elements, as in Fig. 12.5.

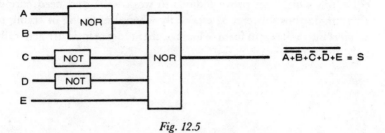

Fig. 12.5

Incidentally a truth table written out for this particular requirement would run to 32 lines, only three of which would give a 'go' output (S = 1)

	A	B	C	D	E	S
line 1 ...	0	0	0	0	0	0
	1	0	0	0	0	0
	.					.
	.					.
	.					.
	1	0	1	1	0	1
	.					.
	.					.
	.					.
	0	1	1	1	0	1
	.					.
	.					.
	.					.
	1	1	1	1	0	1
line 32 ...	1	1	1	1	1	0

Even Boolean algebra, however, has distinct limitations when it comes to more complete circuit designs. It can produce cumbersome equations which can prove tedious to work with, and need careful checking for possible errors, especially when simplifying or trying to convert into a different form of logic. Other types of logic are generally to be preferred in such cases.

CHAPTER 13

Karnaugh Maps and Logic Circuits

A Karnaugh map is a most useful device for combinational circuit design (and can also be used for working other solutions in Boolean algebra). It consists of a square or rectangular area divided into squares, each square representing one minterm. The number of squares required will be 2^N, where N is the number of variables.

Taking just two variables A and B as the simplified example, the corresponding Karnaugh map will have $2^2 = 4$ squares, covering the four possible combinations of A and B (i.e. $\bar{A} \cdot \bar{B}$, $\bar{A} \cdot B$, $A \cdot \bar{B}$, $A \cdot B$).

These squares can be labelled in various ways, two basic methods being shown in Fig. 13.1. That on the left labels the possible states of

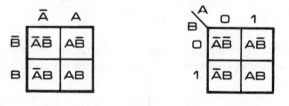

Fig. 13.1

A and B in columns and squares, respectively. That on the left designates A and B allocation – to the right and down from the separating line at the top left corner, with possible signal values. The resulting combinations marked in the squares are shown in minterms, but alternatively could have signal values (i.e. binary 0 or 1). Of the two the left-hand form is probably easiest to use for logic expression reduction (i.e. elimination of redundancies or *minimization*); and the right-hand form probably easiest for translating a truth table into a Karnaugh map.

The sequence in which the variables are presented is not significant, but the squares cannot be allocated at random. They must be arranged so that movement of one square upwards or downwards, or one square left or right horizontally, results in the minterms associated with the two adjacent squares differing only in a single variable. Specifically this implies that the minterms in adjacent squares are identical except for one variable which is marked in one square but not in the other.

Karnaugh maps can accommodate more than two variables. Figs. 13.2 and 13.3 show alternative versions of maps for three and four variables, respectively. This is about the practical limit. Maps with more than five variables are too cumbersome to be of real value.

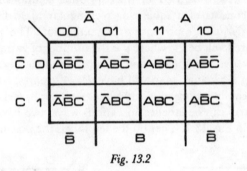

Fig. 13.2

Movement in a Karnaugh map must always be *horizontally* (along squares) or *vertically* (up and down columns), one square at a time. On

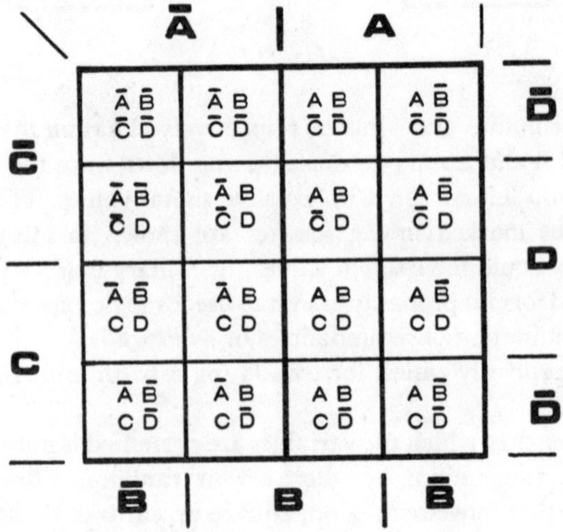

Fig. 13.3

reaching the edge of the map, the next move of one square returns to the opposite edge of that square or column. In other words, a Karnaugh map is basically a three-dimensional device, drawn in two dimensions. How it continues in the third dimension can be understood by thinking of the flat map being rolled up into a vertical cylinder (for continued horizontal movement), or a horizontal cylinder (for continued column movement) – Fig. 13.4.

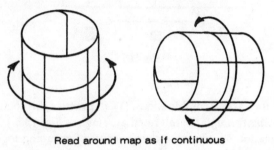

Read around map as if continuous

Fig. 13.4

Minimization

A particular application of a Karnaugh map is for producing a minimal form of Boolean equation. We will use the original equation used in describing minterms as an example:

$$S = A \cdot \bar{B} \cdot \bar{C} + A \cdot B \cdot \bar{C} + \bar{A} \cdot \bar{B} \cdot C + A \cdot B \cdot C$$

The Karnaugh map drawn for this is shown in Fig. 13.5. Fig. 13.6 then shows this map redrawn and re-labelled using the circuit outputs

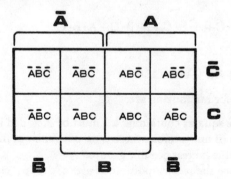

Fig. 13.5

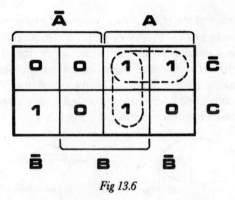

Fig 13.6

relative to the various combinations. (With practice this could be done without going through the intermediate stage of Fig. 13.5.)

Simplification or *minimizing* then consists of grouping together adjacent squares which both contain 1. Grouped squares can then be represented by a single AND term instead of two. There are two such groups in Fig. 13.6, shown enclosed by dashed lines. In the horizontal group, B is the variable that changes and so instead of $A \cdot B \cdot \bar{C} + A \cdot \bar{B} \cdot \bar{C}$ which this group covers, this group can be represented simply by $A \cdot \bar{C}$ (i.e. $A \cdot B \cdot \bar{C} + A \cdot \bar{B} \cdot \bar{C}$ in the original equation becomes $A \cdot \bar{C} (B + \bar{B})$ which $= A \cdot \bar{C}$).

Similarly in the second (vertical) group, $A \cdot B \cdot \bar{C} + A \cdot B \cdot C$, C is the variable that changes, so these two squares can be represented by a single end term $A \cdot B$. There are no other groups and so the original equation:

$$S = A \cdot \bar{B} \cdot \bar{C} + A \cdot B \cdot \bar{C} + \bar{A} \cdot \bar{B} \cdot C + A \cdot B \cdot C$$

reduces to

$$S = A\bar{C} + AB + A \cdot B \cdot C$$

Basic rules for minimizing are:

(i) All the squares containing a 1 must be included in at least one group (if not possible, there is no possibility of minimization).

(ii) Form the largest possible groups.

(iii) Produce the *smallest* possible *number* of groups (consistent with (i) and (ii) above).

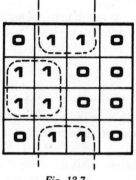

Fig. 13.7

A basic problem in conforming to these rules is that the effect of 'running off the edge' of the map is overlooked, e.g. *see* Fig. 13.7. Here there is already one large group of four squares, and two smaller groups each of two squares at the top and bottom. In fact these two smaller groups form one larger group (think of the map being rolled up into a *horizontal* cylinder). There are, in fact just two groups not three, needing just two AND terms, not three.

Failure to spot how small groups may form a larger group does not matter greatly. It does not upset the validity of the treatment. It simply means that the final answer derived is not as fully simplified as possible. Translated into hardware in a switching circuit, for example, the circuit would work just as well but include more switching or logic components than are strictly necessary.

Sequential Circuits

So far we have only dealt with combinational circuits. Other control circuits may require *sequential* logic, i.e. operations or movements following in a particular sequence. Even more complicated is the *compound* circuit which incorporates both combinational and sequential requirements.

Karnaugh maps can again be used for the design of sequential or compound logic circuits, only where sequence is involved a further type of logic is required – MEMORY for providing the right timing of the sequence. One memory will be required at each sequence point, when *each* memory will require its own Karnaugh map. In a compound circuit, a further Karnaugh map (or maps) would be necessary for the combinational requirements.

Dealing with a typical *sequential* logic circuit, the procedure would be as follows:

  (i)   Write down a word statement of the sequence.
 (ii)   Draw a *time* diagram designating the complete sequence.
(iii)   Prepare a separate *signal flow* diagram and on this plot the signal flow path.
 (iv)   Prepare a Karnaugh map for every memory, and for every auxiliary memory appearing in the signal flow diagram.
  (v)   Extend the presence of memory designations by infilling and forming loops on the signal flow diagram.
 (vi)   Finally, minimize the memory set and reset equations.

No wonder this form of logic is largely incomprehensible to all but specialists!

Even Karnaugh maps have their distinct limitations when it comes to the design of the more complex logic control circuit where, in fact, only design by logic can be relied upon to provide optimal solutions. For that reason many major producers of logic control devices themselves (hardware) have developed their own system methods to make logic designs easier to understand and minimize circuit design time. This will then relate to a specific system which, in turn, is based on a particular choice of logic elements providing all the logic functions necessary.

For a full coverage of all the types of logic required in complex circuits design the functions required are:

NOT AND OR YES NAND NOR INHIBITION
and MEMORY.

That does not mean that all these separate functional *elements* will be required. Solutions can be worked all in AND logic, or all in OR logic (or the inverted forms NAND and NOR). The only functions which remain necessarily common with any type of logic chosen are NOT, INHIBIT and MEMORY. NOT we already know about. INHIBITION is a sort of variation on NOT and AND, expressed as:

B and *not* A = S
or B·Ā = S
B·A = 0

In other words, when a signal is present at A it *inhibits* the signal path of B.

MEMORY is readily performed by a flip-flop device accepting two signal inputs A and B. A signal applied at input A will give an output state S1 and maintain this state even when signal A is removed. A signal input at B will then change the state of the memory to S2, which it will continue to hold when signal B is removed, until tripped back to state S1 when a further signal is applied to A. In other words, MEMORY is a logic device which 'remembers' the last signal, which is an essential feature in sequential logic.

As far as logic circuit components are concerned (i.e. the available 'hardware' for turning a circuit design into a working circuit), choice of *type* of logic used depends on the suitability of such 'working' elements to perform the various functions necessary. This is no real problem in electronic circuits using solid state devices, but in the case of pneumatic and hydraulic circuits, the working elements involved are normally miniaturized valves. If a multiplicity of individual elements (valves) is to be avoided, the choice favours using NOR or NAND logic, as these functions can be performed by simpler valves. Other systems may, however, employ multi-functional devices based on OR, AND, NOT, YES and MEMORY.

Index

Other Bestsellers From TAB